Lync Server Cookbook

Over 90 recipes to empower you to configure, integrate, and manage your very own Lync Server deployment

Fabrizio Volpe

Alessio Giombini

Lasse Nordvik Wedø

António Vargas

BIRMINGHAM - MUMBAI

Lync Server Cookbook

Copyright © 2015 Packt Publishing

All rights reserved. No part of this book may be reproduced, stored in a retrieval system, or transmitted in any form or by any means, without the prior written permission of the publisher, except in the case of brief quotations embedded in critical articles or reviews.

Every effort has been made in the preparation of this book to ensure the accuracy of the information presented. However, the information contained in this book is sold without warranty, either express or implied. Neither the authors, nor Packt Publishing, and its dealers and distributors will be held liable for any damages caused or alleged to be caused directly or indirectly by this book.

Packt Publishing has endeavored to provide trademark information about all of the companies and products mentioned in this book by the appropriate use of capitals. However, Packt Publishing cannot guarantee the accuracy of this information.

First published: January 2015

Production reference: 1210115

Published by Packt Publishing Ltd.
Livery Place
35 Livery Street
Birmingham B3 2PB, UK.

ISBN 978-1-78217-347-2

www.packtpub.com

Credits

Authors
Fabrizio Volpe
Alessio Giombini
Lasse Nordvik Wedø
António Vargas

Reviewers
Pantelis Apostolidis
Gianluca Bellu
Tonino Bruno
Randy Chapman
Desmond LEE
Clinton Mann
Johan Veldhuis

Commissioning Editor
Taron Pereira

Acquisition Editor
Kevin Colaco

Content Development Editor
Neeshma Ramakrishnan

Technical Editor
Indrajit A. Das

Copy Editors
Karuna Narayanan
Alfida Paiva

Project Coordinator
Danuta Jones

Proofreaders
Simran Bhogal
Maria Gould
Ameesha Green
Paul Hindle

Indexer
Priya Sane

Production Coordinator
Shantanu N. Zagade

Cover Work
Shantanu N. Zagade

About the Authors

Fabrizio Volpe is a Lync MVP and an experienced IT professional, with more than 15 years of experience working in the IT department of large-scale banking and financial companies. He has been working as a network and systems administrator in various firms of the Iccrea Banking Group (one of the top banking groups in Italy) since 2000. From 2011 to 2013, before moving his focus to Unified Communications, Fabrizio received the MVP award for Directory Services.

Over the past few years, Fabrizio has participated as a speaker at many events and conferences (both Italian and international). He creates IT-focused content on different platforms that have received good feedback. His works are available on YouTube (http://www.youtube.com/user/lync2013), on his personal blog (http://www.absoluteuc.org), and on SlideShare (http://www.slideshare.net/fabriziov).

Fabrizio has published three books with Packt Publishing: *Getting Started with FortiGate*, *Getting Started with Microsoft Lync Server 2013*, and *Instant Microsoft Forefront UAG Mobile Configuration Starter*. He has also authored a successful free e-book, *Microsoft Lync Server 2013: Basic Administration*, available in the TechNet Office gallery (http://bit.ly/1jbzpfo), which has been downloaded more than 25,000 times.

Questo libro è dedicato a mia moglie, Antonella.

Alessio Giombini is a solutions architect, with a strong focus on Microsoft and Unified Communications area. He has over 15 years' worth of study and hands-on experience delivering small- to large-scale projects for major enterprise industries, mainly based on Microsoft and leading-edge technologies, systems applications, and operations running on top of them. He has a broad and mixed technical background in the IT infrastructure and communications field, systems integration, systems management, security, as well as an in-depth understanding of the business of computing and networking in enterprise organizations.

Alessio Giombini currently works for Intercall EMEA, based in the United Kingdom, as a solutions architect for Unified Communications platforms. He designs and deploys UC infrastructures based on Microsoft Lync and related technologies, including private and public cloud-based hosted solutions on multitenanted and dedicated Lync architectures. He also loves talking about Lync through presales, by delivering technical presentations and workshops, solution designs, writing HLD and LLD documents, delivering proof of concepts, and designing the solution through to implementation.

> I wish to deeply thank my family, the lights of my life, Roberta, Carlotta, Giorgio and Nico, for their love and patience.

Lasse Nordvik Wedø has more than 14 years of experience working with large-scale IT infrastructure, specializing in planning, deploying, and supporting Unified Communications systems from both Microsoft and Cisco.

He previously specialized in planning, deploying, and supporting Windows Active Directory solutions, where security and messaging were his main areas of focus. He has recently been made a Microsoft P-TSP in Norway. In his spare time, he contributes to the Lync community through his own blog (http://tech.rundtomrundt.com), where he likes to share his thoughts and helpful scripts. He is also a contributor to a blog dedicated to helping admins organize their Enterprise Voice number plans (http://lyncnumbers.com). He was a speaker at Norwegian Lync day 2014 and TechEd Europe 2014.

António Vargas is a Microsoft Certified Solutions Master in Exchange 2013 with 15 years of experience as a Microsoft consultant, designing and deploying large-scale projects for customers across all industry sectors.

The main focus of his Microsoft projects is on the Unified Communications portfolio, more specifically on Microsoft Exchange, Microsoft Office 365, and Microsoft Lync.

Currently, Antonio works for Intercall EMEA, based in the United Kingdom, as a Microsoft Unified Communications architect, planning, designing, and delivering migrations and greenfield deployments of Microsoft Exchange and Microsoft Office 365 environments. Most of his work also includes configuring all levels of integration between Microsoft Lync and Microsoft Exchange, on premises or on Office 365.

About the Reviewers

Pantelis Apostolidis is an IT professional who is passionate and has been working for a decade with almost all Microsoft IT Pro services, including the Domain, Exchange, Lync, System Center, Office 365, and Azure. His educational background includes a diploma in tourism management from the Technological Educational Institute (TEI) of Thessaloniki, and a diploma in computer network engineering from the IEK of Thessaloniki. He is a Microsoft Certified Solutions Expert (MCSE) Private Cloud. He has other certifications, which include MCSA 2012, MCITP EMA 2010, MCTS, MBSS, MS, MCSA 2003, and MCP. During his free time, he enjoys playing the guitar and reading fiction, but most of all, he enjoys spending time with his wife and two kids. He also blogs at http://proximagr.wordpress.com.

Gianluca Bellu started his IT career in 2001 in Rome (Italy) as a system engineer and a developer, and focused on Microsoft infrastructures for Line of Business applications at Nextiraone, Italy, which is a system integrator company with its presence in 15 countries.

In 2004, he was an IVR developer and a call center engineer at Alcatel Technologies. From 2006, he started working on Microsoft Office Communications Server 2007 (as it was in the beta version), and he was responsible for the business proposals and presales/postsales activities based on Microsoft UC as a new and innovative technology.

In 2014, he was hired by BT Switzerland Ltd. as a UCC Specialist in the BT Advise team based in Zurich.

Gianluca Bellu is now a Microsoft Certified IT Professional (MCITP) on Lync Server and a Microsoft Certified Application Developer (MCAD); during his career, he has also achieved Cisco CCNA certification, Snom certification (SCE), and many more on Audiocodes, Cycos, and Genesys.

He's the owner and blogger at http://msucblog.wordpress.com, in which he shares his know-how on Microsoft Unified Communications.

Tonino Bruno is a Microsoft Certified Master in messaging and a subject-matter expert on large-scale, complex, and cross-premises messaging solutions based on Microsoft Exchange and Lync Server. With over 13 years of experience as a subject matter expert, Tonino has become a trusted advisor for many of Belgium's largest and even international corporations. After having worked for 9 years at Compaq/HP, Tonino has successfully launched his own consulting firm and has been working in close collaboration with Microsoft Services in Belgium and Luxembourg for the past 5 years.

Randy Chapman is a Lync architect, evangelist, and blogger. Randy has worked with computers since the early 80s. Randy has done everything from sales and support, demonstration and design, installation, and training of everything from PCs and servers to communications systems for very small to very large companies. He has worked for over 15 years as a voice consultant and over 10 years as a Microsoft consultant with good knowledge, spanning the entire Microsoft stack. He has worked extensively with Exchange Server since 2000 and with System Center since 2008. He has been working with Microsoft-based Unified Communications for over 5 years. Over the years, he has designed, delivered, and supported many deployments of Lync 2010 and 2013. He has also integrated and replaced many PBXs from vendors such as Cisco, Avaya, Mitel, Alcatel, Ericsson, and Bosch to name a few, and provided training for end users, sales and support staff, as well as the next generation of UC consultants. He is a competent and confident speaker, who has spoken and demonstrated at Love Lync Events throughout the UK. Randy is active on many forums, including LinkedIn and TechNet, and his contribution to the TechNet Gallery has been downloaded thousands of times. Randy's Lync blog, `http://lynciverse.com`, gets more than 1,800 visitors a month.

Randy works for MeetingZone as a senior Lync architect and subject-matter expert. He helps companies all over the UK visualize and realize the benefits of Microsoft Lync-based UC solutions.

> I would like to thank my wife, Tammy, for allowing me to continue to feed my obsession with all things Lync and Unified Communications. I would like to thank Simon, Stuart, and James for my start and for your help and support over the years. I would also like to thank all of the great Lync pros out there for fueling my passion. Finally, a special thanks to Fabrizio Volpe for giving me the opportunity to work on a project I feel so passionately about.

Desmond LEE specializes in end-to-end enterprise infrastructure and cloud solutions built around proven business processes and people integration across various industries. He is recognized as a Microsoft Most Valuable Professional (MVP Lync Server) for his passion and volunteer work in the IT community. Desmond is a long-time Microsoft Certified Trainer (MCT) and founder of the Swiss IT Pro User Group (www.swissitpro.com), an independent, non-profit organization for IT pros by IT pros championing Microsoft technologies.

An established speaker at major international and regional events and known for his real-world insights, Desmond contributes frequently to several highly rated publications, and acts as a moderator in popular Microsoft public forums/newsgroups. You can follow his IT adventures at www.leedesmond.com.

Clinton Mann has over 16 years of professional experience in the field of IT, and over those 16 years, he has worn many hats, and man, does he like his hats. With each of the hats he has worn, Clinton has always made it a point to learn and share new technologies with anyone who would listen, and most people benefit from his expertise. Clinton has a technolust like none before him; he truly does live and breathe technology.

Clinton currently works as an IT systems engineer at the Wyss Institute for Biologically Inspired Engineering at Harvard University. You can also find him living the digital dream on the Internet, of all the places, at https://www.linkedin.com/in/clintonmann. You can find him on Twitter at @manncl, and his website can be found at http://clintonmann.com.

> I would like to thank Packt Publishing for giving me the opportunity to check an item off my bucket list: *Review a book*. I would also like to express my deepest gratitude to my wife, who is the *North Star* in my life and who always provides me with her light, support, and guidance. I would also like to thank my own wolf pack, Hendrix, Rizzio, and Rams.

Johan Veldhuis works as a Premier Field Engineer (PFE) for Microsoft. In his current role, he delivers both proactive and reactive services to customers from Microsoft for both Microsoft Exchange and Lync.

In his spare time, Johan spends time blogging via his own website (www.johanveldhuis.nl). Besides blogging, Johan is a member of the UC Architects (www.theucarchitects.com), which is a biweekly podcast, where Exchange and Lync freaks discuss both Exchange- and Lync-related topics.

www.PacktPub.com

Support files, eBooks, discount offers, and more

For support files and downloads related to your book, please visit www.PacktPub.com.

Did you know that Packt offers eBook versions of every book published, with PDF and ePub files available? You can upgrade to the eBook version at www.PacktPub.com and as a print book customer, you are entitled to a discount on the eBook copy. Get in touch with us at service@packtpub.com for more details.

At www.PacktPub.com, you can also read a collection of free technical articles, sign up for a range of free newsletters and receive exclusive discounts and offers on Packt books and eBooks.

https://www2.packtpub.com/books/subscription/packtlib

Do you need instant solutions to your IT questions? PacktLib is Packt's online digital book library. Here, you can search, access, and read Packt's entire library of books.

Why subscribe?

- Fully searchable across every book published by Packt
- Copy and paste, print, and bookmark content
- On demand and accessible via a web browser

Free access for Packt account holders

If you have an account with Packt at www.PacktPub.com, you can use this to access PacktLib today and view 9 entirely free books. Simply use your login credentials for immediate access.

Instant updates on new Packt books

Get notified! Find out when new books are published by following @PacktEnterprise on Twitter or the *Packt Enterprise* Facebook page.

Table of Contents

Preface   **1**

Chapter 1: Lync 2013 Security   **7**
- Introduction   7
- Controlling administrative rights with RBAC and custom cmdlets   8
- Hardening Lync Servers   10
- Hardening Lync databases   14
- Enhancing conferencing security   18
- Managing certificates for the authentication of desk-phones   19
- Deploying a secure Lync Edge   21
- Applying ethical walls for federation security   23
- Using Application Request Routing to configure a reverse proxy for Lync Server 2013   26

Chapter 2: Lync 2013 Authentication   **33**
- Introduction   33
- Configuring passive authentication for Lync   35
- Enabling two-factor authentication   41
- Adding the app password for mobile clients   45
- Authenticating with online services using DirSync   46
- Managing Windows Azure Directory for Lync Online   54
- Configuring server-to-server authentication   56
- Troubleshooting with client authentication logging   62

Chapter 3: Lync Dial Plans and Voice Routing   **63**
- Introduction   63
- Introducing dial plans and voice routing   64
- Defining dial plans   69
- Configuring PSTN usage – voice policy   72
- Configuring PSTN usage – Location-Based Routing   76
- Enabling routes   78

Validating trunks	82
Configuring load balancing, failover, and least cost routing	83
Controlling call forwarding	87

Chapter 4: Lync 2013 Integration with Exchange — 89

Introduction	89
Configuring the Unified Messaging integration	93
Configuring OAuth between Lync 2013 and Exchange 2013	105
Configuring Lync 2013 and Exchange 2013 as partner applications	107
Configuring Lync 2013 to use Exchange 2013 for archiving	109
Configuring Lync 2013 to use the Exchange 2013 Unified Contact Store	112
Integrating Lync 2013 with the Exchange 2013 Outlook Web App	114

Chapter 5: Scripts and Tools for Lync — 119

Introduction	119
Installing Lync prerequisites and more – Set-Cs2013Features	120
Creating a fully functional voice configuration – Lync Dialing Rule Optimizer	123
Switching between multiple Lync identities with a click – Profiles for Lync (P4L)	127
Tracing made easier – Lync 2013 Centralized Logging Tool	128
Identifying recurrent issues – Lync Pilot Deployment Health Analysis	132
Managing phone numbers – Search-LineURI and Get-UnusedNumbers	138
Managing Call Pickup Groups – Lync2013CallPickupManager 1.01	143

Chapter 6: Designing a Lync Solution – The Overlooked Aspects — 147

Introduction	147
Meeting your users' expectations	148
User training	151
Gathering the users' requirements	153
Weighing up around Lync virtualization	155
Network readiness – introduction	159
Defining personas for the network	162
Defining sites for the network	164
Network readiness – reviewing and analyzing results	167

Chapter 7: Lync 2013 in a Resource Forest — 173

Introduction	173
Planning a resource forest	174
Using Exchange Online for a Lync resource forest	184
Configuring FIM in a Lync resource forest	187
Synchronizing forests with FIM	194
Deploying Azure Active Directory	198

Synchronization services (AAD Sync) in a Lync resource forest	198
AAD Sync synchronization services and rules	202
Chapter 8: Managing Lync 2013 Hybrid and Lync Online	**207**
Introducing Lync Online	207
Administering with the Lync Admin Center	208
Using Lync Online Remote PowerShell	211
Using Lync Online cmdlets	215
Introducing Lync in a hybrid scenario	218
Planning and configuring a hybrid deployment	220
Moving users to the cloud	222
Moving users back on-premises	225
Debugging Lync Online issues	226
Chapter 9: Lync 2013 Monitoring and Reporting	**233**
Introduction	233
Installing Lync 2013 monitoring reports	234
Selecting the right kind of report	239
Call Diagnostic Reports	244
Media Quality Diagnostic Reports	248
Call Leg Media Quality Report	252
Lync 2013 with System Center 2012 R2 Operations Manager	254
Configuring a watcher node and synthetic transactions	260
Chapter 10: Managing Lync 2013 Backup and Restore	**263**
Introduction	263
Topology information	264
Configuration information	270
User database	274
Persistent Chat database	277
The Location Information LIS database	279
The Response Group Services configuration	282
Certificates	284
Backend databases	290
Voice dial plans, policies, and settings	292
File services	296
Don't forget the infrastructure – the greater recovery plan	299
Chapter 11: Controlling Your Network – A Quick Drill into QoS and CAC	**303**
Introduction	304
Gathering data about your network	307
Creating network bandwidth policies	312

Adding networks to the topology	315
Creating region links and routes	320
Enabling CAC	324
Preparing servers and clients for DSCP tagging	327
Controlling/limiting the port ranges for traffic	330
Media bypass	333
Chapter 12: Lync 2013 Debugging	**335**
Introduction	335
Using Snooper to examine log files	336
Investigating Call Flow with Snooper Flow Chart	340
Reviewing Lync information with OCSLogger	342
Tracing from a command line with OCSTracer	346
Customizing CLS scenarios using CLSController	347
Testing our setting with Best Practices Analyzer	349
Capturing network traffic with Wireshark	351
Troubleshooting clients with the Microsoft Lync Connectivity Analyzer	359
Verifying a deployment with the Microsoft Remote Connectivity Analyzer	362
Index	**365**

Preface

This is the first cookbook dedicated to Lync Server 2013. While there are a few books dedicated to Lync (and Packt Publishing has published a couple of them), this is the first time that someone has tried to write down more than 300 pages of practical recipes, hints, and tips dedicated to this Unified Communication software.

While writing this book, we, as authors, know that Skype for Business will be available during this year, and a part of the existing features and interfaces could change to some extent. We are confident, anyway, that Lync Server 2013 will stay relevant for a long time, and we believe that people working on solutions based on Lync around the world will value the time and effort we have put into this cookbook.

We have tried to include in this book as much useful information as possible, to help Lync administrators in their everyday tasks and in planning, deploying, and managing some of the most complex scenarios and features. The coming years will see an increase in the use of cloud computing, and Lync, right now, integrates in many ways with Azure and Office 365. We have tried to explain the cloud-related features and options, in addition to the more traditional on-premises settings. As authors, we accept the risk that the ever-changing nature of the Microsoft Cloud might require updates to the material in this book, because our commitment is to provide you with the most relevant information.

As we said, this book is something new both in terms of format and content. We hope it will be like a tool that you will keep on your "desk", and consult over and over as the need arises.

What this book covers

Chapter 1, Lync 2013 Security, is dedicated to the hardening techniques for the Lync infrastructure and to some recipes to raise the level of security for some of the available features. This chapter includes a configuration guide to use Application Request Routing as a reverse proxy for Lync.

Chapter 2, *Lync 2013 Authentication*, focuses on the authentication protocols used in Lync for the various devices and identities that have access to the server features. This chapter contains recipes dedicated to authentication configuration and management, both on-premises and on the cloud.

Chapter 3, *Lync Dial Plans and Voice Routing*, discusses Enterprise Voice, which is the most complex feature to plan and administer in Lync Server 2013. Although a complete overview of such a vast topic is not possible, this chapter focuses on the management of dial plans and voice routing, introducing a series of real-world suggestions and recipes.

Chapter 4, *Lync 2013 Integration with Exchange*, requires comprehension of Exchange to deliver features such as Unified Messaging integration and Lync archiving with Exchange and the Unified Contact Store. The recipes in this chapter will help you in the tasks related to Lync / Exchange integration.

Chapter 5, *Scripts and Tools for Lync*, contains an overview of useful tools that every Lync administrator should know. The software and scripts presented in this chapter are so important that, in some cases, we have used them in a more extensive manner in other feature-focused chapters.

Chapter 6, *Designing a Lync Solution – The Overlooked Aspects*, takes care of some aspects that are often ignored during the design phase of a Lync solution. The human factor (such as training and assessment of user requirements) and more technical aspects are examined in this chapter.

Chapter 7, *Lync 2013 in a Resource Forest*, explores the different solutions available to maximize our Lync deployment with the use of a resource forest. The scenarios proposed include both on-premises and hybrid solutions to deliver Lync features to the users' forests.

Chapter 8, *Managing Lync 2013 Hybrid and Lync Online*, gives an overview of the tools and techniques required to manage Lync Online and to administer a hybrid deployment of Lync. There are recipes dedicated to help Lync administrators perform the most common administrative tasks in the previously mentioned scenarios.

Chapter 9, *Lync 2013 Monitoring and Reporting*, covers the concept of monitoring, which is a crucial aspect of a Lync production environment. Lync offers some default reports to monitor the health of our deployment and the quality of the audio and video experience that we offer to our users. The recipes in this chapter are dedicated both to the use of the previously mentioned information and to the configuration of additional controls.

Chapter 10, *Managing Lync 2013 Backup and Restore*, covers the Lync architecture, which contains mechanisms that grant a high level of continuity. Anyway, we have to provide a consistent plan to prevent data loss and configuration corruptions. This chapter is focused on identifying the information that we need to back up and explaining the ways to restore our working environment.

Chapter 11, Controlling Your Network – A Quick Drill into QoS and CAC, grants the best experience to our users, which is one of the most important aspects in every Lync deployment. Delivering audio and video services with no use of quality of service and no call control is a risky decision, which usually leads to offering services with a low level of performance. We have some important recipes in this chapter that cover the configuration and use of QOS and CAC in Lync Server 2013.

Chapter 12, Lync 2013 Debugging, discusses the concept of troubleshooting in Lync, which is usually a complicated task. This last chapter of the book lists and explains some of the best tools available to resolve different kinds of problems that we could face in a Lync environment.

What you need for this book

To deploy Lync, the list of required software (also counting the additional software) includes the following:

- Lync Server 2013 Standard or Enterprise edition
- A compatible operating system (Windows Server 2008 R2 SP1, Windows Server 2012, or Windows Server 2012 R2)
- Microsoft SQL Server 2008 R2 Enterprise or Standard, or SQL Server 2012 Enterprise or Standard, is required for Lync Enterprise Edition and for all the databases that we want to keep separated from the local installation of SQL Express that is part of any Lync Front End
- Office Web Apps

Who this book is for

This book is dedicated to Lync administrators, no matter the size of their deployment. People involved in a Lync project can use the book both for a specific recipe or to have an overview of some specific scenarios and configurations.

Sections

In this book, you will find several headings that appear frequently (Getting ready, How to do it, How it works, There's more, and See also).

Preface

To give clear instructions on how to complete a recipe, we use these sections as follows:

Getting ready

This section tells you what to expect in the recipe, and describes how to set up any software or any preliminary settings required for the recipe.

How to do it...

This section contains the steps required to follow the recipe.

How it works...

This section usually consists of a detailed explanation of what happened in the previous section.

There's more...

This section consists of additional information about the recipe in order to make the reader more knowledgeable about the recipe.

See also

This section provides helpful links to other useful information for the recipe.

Conventions

In this book, you will find a number of text styles that distinguish between different kinds of information. Here are some examples of these styles and an explanation of their meaning.

Code words in text, database table names, folder names, filenames, file extensions, pathnames, dummy URLs, user input, and Twitter handles are shown as follows: "In our example, we have the resource forest (`wonderland.lab`) and the user forest (`forest.lab`)."

A block of code is set as follows:

```
New-MsolServicePrincipalCredential -AppPrincipalId 00000002-0000-0ff1-
ce00-000000000000 -Type Asymmetric -Usage Verify -Value
$credentialsValue -StartDate 7/15/2014 -EndDate 7/3/2015
```

Any command-line input or output is written as follows:

```
Set-CsVoicePolicy UK-London-Local -Allowcallforwarding $false
```

New terms and **important words** are shown in bold. Words that you see on the screen, for example, in menus or dialog boxes, appear in the text like this: "The **UK-London-Local** policy allows all forwarding."

> Warnings or important notes appear in a box like this.

> Tips and tricks appear like this.

Reader feedback

Feedback from our readers is always welcome. Let us know what you think about this book—what you liked or disliked. Reader feedback is important for us as it helps us develop titles that you will really get the most out of.

To send us general feedback, simply e-mail `feedback@packtpub.com`, and mention the book's title in the subject of your message.

If there is a topic that you have expertise in and you are interested in either writing or contributing to a book, see our author guide at `www.packtpub.com/authors`.

Customer support

Now that you are the proud owner of a Packt book, we have a number of things to help you to get the most from your purchase.

Downloading the example code

You can download the example code files from your account at `http://www.packtpub.com` for all the Packt Publishing books you have purchased. If you purchased this book elsewhere, you can visit `http://www.packtpub.com/support` and register to have the files e-mailed directly to you.

Errata

Although we have taken every care to ensure the accuracy of our content, mistakes do happen. If you find a mistake in one of our books—maybe a mistake in the text or the code—we would be grateful if you could report this to us. By doing so, you can save other readers from frustration and help us improve subsequent versions of this book. If you find any errata, please report them by visiting http://www.packtpub.com/submit-errata, selecting your book, clicking on the **Errata Submission Form** link, and entering the details of your errata. Once your errata are verified, your submission will be accepted and the errata will be uploaded to our website or added to any list of existing errata under the Errata section of that title.

To view the previously submitted errata, go to https://www.packtpub.com/books/content/support and enter the name of the book in the search field. The required information will appear under the **Errata** section.

Piracy

Piracy of copyrighted material on the Internet is an ongoing problem across all media. At Packt, we take the protection of our copyright and licenses very seriously. If you come across any illegal copies of our works in any form on the Internet, please provide us with the location address or website name immediately so that we can pursue a remedy.

Please contact us at copyright@packtpub.com with a link to the suspected pirated material.

We appreciate your help in protecting our authors and our ability to bring you valuable content.

Questions

If you have a problem with any aspect of this book, you can contact us at questions@packtpub.com, and we will do our best to address the problem.

1
Lync 2013 Security

In this chapter, we will cover the following topics:

- Controlling administrative rights with RBAC and custom cmdlets
- Hardening Lync Servers
- Hardening Lync databases
- Enhancing conferencing security
- Managing certificates for desk-phones authentication
- Deploying a secure Lync Edge
- Applying ethical walls for federation security
- Using Application Request Routing to configure a reverse proxy for Lync Server 2013

Introduction

There is a high level of security inherent in all the Lync Server features. Unified communications, from a customer's point of view, require a special level of privacy and control, and Lync is designed with mechanisms to answer to this need in a clear manner. Lync updates (both on the client and on the server side) have added to the software a flexibility in design, so that it is now possible, for example, to use certificate authentication or passive authentication for mobility scenarios, or to add a two-factor authentication (as we will see in *Chapter 2, Lync 2013 Authentication*). In this chapter, we are going to talk about some of the security aspects related to the infrastructure. Lync 2013 security has two different scopes, one related to the network where the servers are located, and one related to the services we make available to the external users. The recipes regarding Role-Based Access Control and servers and database hardening are more relevant to protect our deployment from threats that come from the corporate network, while the topics related to ethical walls, reverse proxy, and edge security are fundamental aspects when the communication extends to the Internet.

A fundamental document that you should read as a starting point is the *Security Framework for Lync Server 2013* post (http://technet.microsoft.com/en-us/library/dn481316.aspx), which will give you a high-level overview of the security features inside Lync Server 2013.

Controlling administrative rights with RBAC and custom cmdlets

Lync Server 2013 administration uses **Role-Based Access Control** (**RBAC**) to assign different levels of access privileges to the users, and to enable them to perform specific administrative tasks. The idea behind RBAC in Lync 2013 is that adding a user to a specific group not only defines the features and administrative tasks they are able to manage but also limits the cmdlets they are able to use in the Lync Management Shell. There are some built-in administrative roles, and we are able to add custom groups for more granular control. Another operation we are able to perform is adding authorized cmdlets to both kinds of groups, expanding the allowed tasks for a specific RBAC role.

Getting ready

In our example, we will use both of the previously mentioned customizations, creating a new customized user group, CsUserModifier, based on the default group CsViewOnlyAdministrator, and adding access to the Set-CsUser cmdlet (to modify properties for existing user accounts).

How to do it...

1. Create the CSUserModifier user group (with the scope as universal and type as security) in Active Directory.

2. Open the Lync Server Management Shell and launch the following cmdlet:

   ```
   New-CsAdminRole -Identity CsUserModifier -Template CsViewOnlyAdministrator
   ```

   The cmdlet will clone the permissions of the CsViewOnlyAdministrator group to the custom group.

3. Launch the following cmdlet to verify the list of administrative tasks delegated to the new group:

   ```
   Get-CsAdminRole CSUserModifier | Select-Object -ExpandProperty cmdlets | fl
   ```

The output will be similar to what is shown in the following screenshot:

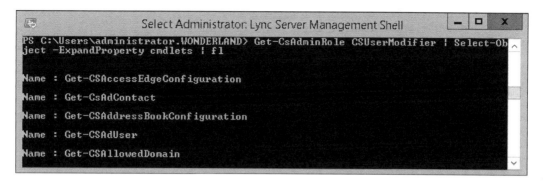

4. Now, we are able to use the cmdlet customization, adding the `Set-CsUser` cmdlet to the available tasks:

   ```
   Set-CsAdminRole -Identity CsUserModifier -Cmdlets @{add="Set-CsUser"}
   ```

5. The same command, with an `@{remove` parameter, can be used to remove some administrative tasks that were previously available from a group:

   ```
   Set-CsAdminRole -Identity CsUserModifier -Cmdlets @{remove="Get-CSVoiceRoutingPolicy","Get-CSVoiceTestConfiguration"}
   ```

6. Verification of the previously mentioned cmdlet is done using the same process we used in step 2, to verify the list of delegated tasks.

7. The `New-CSAdminRole` cmdlet supports the `-Cmdlets` switch that we saw in step 5, so when defining a custom group role, we are able to add custom cmdlets. A command like the next one could achieve both role customization and cmdlet customization in a single step:

   ```
   New-CsAdminRole -Identity CSUserModifier -Template CsViewOnlyAdministrator -Cmdlets @{add="set-CsUser"}
   ```

There's more...

As important as it is for security, RBAC has a severe limitation because it is effective only for users that are working with Lync administrative tools from a remote workstation (http://technet.microsoft.com/en-us/library/gg425917.aspx). The controls are not enforced for users who are working locally on the Lync Server (or using a remote PowerShell session). Physical security of our servers is an important topic, and we should address it with all the available solutions (smart card access, doors, cameras, strong passwords, lights-out servers with no physical keyboard or monitor available, and so on).

Hardening Lync Servers

Talking about Lync Server 2013, we are interested in applying a defense-in-depth approach, using multiple defense layers against security threats. Various security solutions are applied to make bypassing of one of the layers more difficult. We are also able (at least) to buy time on the different layers before someone is able to access the next level of security. Our servers are the last layer before internal data and files of Lync are compromised. Hardening a Lync Server requires a series of steps, and we will see how to use the **Security Configuration Wizard** (**SCW**), a tool that makes it easier to fix some common misconfigurations and security flaws.

Getting ready

To increase the security of the operating system, we can use the SCW (if we are using Windows 2012 or Windows 2012 R2 SCW it is an integrated tool). In the previously mentioned OS, the **Configuration Wizard** is part of the **Tools** menu.

> While the following steps have been tested on a single installation Front End (Lync Server 2013 Standard Edition), we have to select the settings that best fit our specific security requirements, and verify them in a lab. Using SCW on a production environment without sufficient verification is a risky approach.

How to do it...

1. The **Security Configuration Wizard** option is accessible from the **Tools** menu in **Server Manager**, as we can see in the following screenshot:

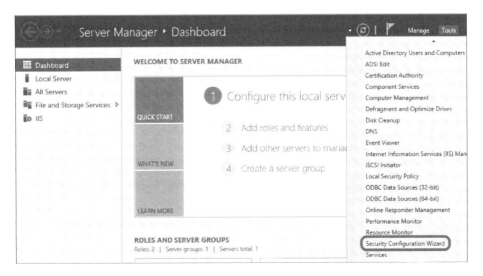

2. In the first screen, we have to click on **Next** and then on **Create a new security policy**.
3. Select the name of the Lync Server that will act as a baseline (in our scenario, Madhatter).
4. We can select **Next** in the **Role-Based Service configuration** screen and again **Next** in the **Select Server Roles** screen until we arrive at **Select client features**. Flag all the options and select **Next**.
5. In the **Select Administration and Other Options** screen, select **Background Intelligent Transfer Service BITs** and **Windows Audio**.
6. In the **Select Additional Services** menu, flag all the services and click on **Next**.
7. Select **Do not change the startup mode of the service** and then select **Next**.
8. In the C**onfirm Service Change** screen, accept the default value and select **Next**.
9. Flag the **Skip this section in the Network Security** screen (we have to leave the flag box clear if we want to use the Windows firewall too).
10. Click on **Next** in the **Registry Settings** screen (if we don't have a legacy operating system that needs to connect to the Lync Server).
11. In the **Require SMB Security Signatures** screen, we can clear the second flag as shown in the following screenshot. If the server has enough unused processing resources, digital signature is a security option that we can consider:

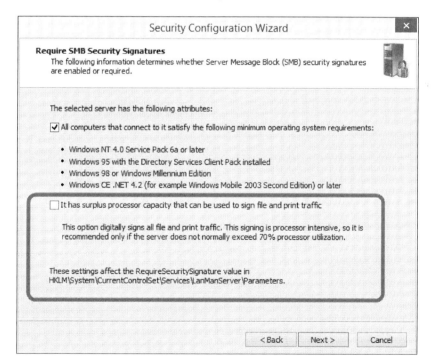

12. As we can see in the following screenshot, in the **Outbound Authentication Methods** screen, we can leave the default settings as is if we are not going to use local accounts to authenticate with other servers:

Outbound Authentication Methods
The following information is used to determine the LAN Manager authentication level used when making outbound connections.

Select the methods the selected server uses to authenticate with remote computers:
- [✓] Domain Accounts
- [ ] Local Accounts on the remote computers
- [ ] File sharing passwords on Windows 95, Windows 98, or Windows Millennium Edition

These settings affect the lmcompatibilitylevel value in HKLM\System\CurrentControlSet\Control\LSA.

13. In the **Outbound Authentication using Domain Accounts** screen, we can leave the default settings as is and select **Next** with no changes also in **Registry Settings Summary**.
14. If we are not going to use auditing, we can select **Skip this section** in the **Audit Policy** screen, and then click on **Next** in the **Save Security Policy** screen.
15. Type a name for the policy (for example, **LyncTemplate**), optionally adding a description, and click on **Next**.
16. Select **Apply Now**, and then click on **Next** and **Finish** in the **Completing the Security Configuration Wizard** screen.
17. It is advisable to reboot the server in order to verify that the new settings have impacted the Lync Server startup phase.

How it works...

If any issue arises with the SCW, we are able to roll back to the previous configuration. If we don't have access to the local server, we can launch the SCW on another server and revert to the configuration remotely. The option is the one we can see in the following screenshot:

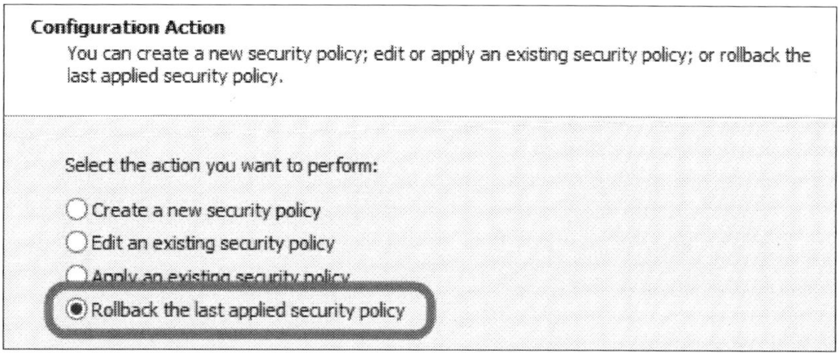

There's more...

SCW can close TCP ports `8080` and `4443` on the Lync Front End. Running the `Enable-CsComputer` cmdlet, we are able to open again the required ports on the Windows Firewall. The same result can be obtained by using **Lync Server Deployment Wizard** or **Bootstrapper.exe**. For more details, see *Re-activate server after Security Configuration Wizard closes ports in IIS* (http://technet.microsoft.com/en-us/library/gg398851.aspx).

SCW can disable the RDP access. We are able to restore the feature with various solutions, for example, by selecting **Remote Desktop** from the **Installed options** list in the **Select Administration and Other Options** screen, as we can see in the following screenshot:

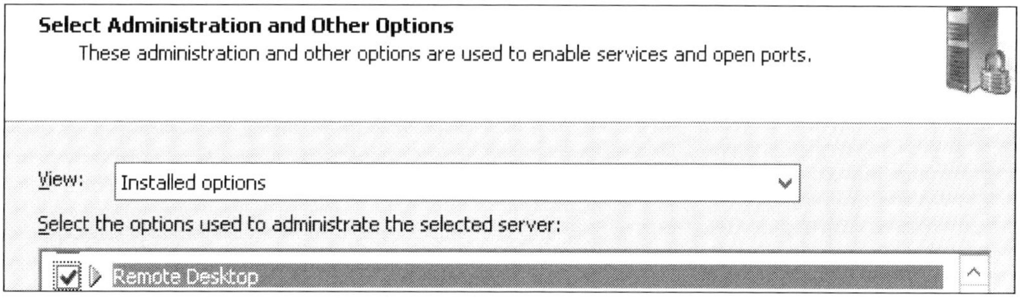

See also

One of the obvious steps to enhance server security is the installation of an antivirus application. To avoid issues with Lync, we should follow the guidelines in this post *Antivirus scanning exclusions for Lync Server 2013* post at http://technet.microsoft.com/en-us/library/dn440138.aspx.

Hardening Lync databases

Lync Server 2013 uses SQL Server as a repository for key information such as the **Central Management Store** (**CMS**), which contains our Lync topology. Lync Standard Edition uses a collocated SQL Server Express backend database that we are not able to move on a different server. Although this configuration reduces the number of machines required for the Lync Server setup, this also limits the options we have to protect our databases. The suggestions in the *There's more...* and *See also* sections are usable for both the Standard Edition and Enterprise Edition of Lync Server. The steps in the *How to do it...* section are applicable only to Lync Server 2013 Enterprise Edition, which has a configuration based on SQL Server that runs on a separate server (with cluster and mirroring supported as a continuity solution).

There are different ways to protect a SQL server, including security measures for the filesystem and best practices, which we will see after the *How to do it...* section. The steps we will see now are meant to make it more difficult to attack our SQL database from the network. SQL server uses a standard port (TCP `1433`) for the default database instance, and TCP `1434` for the SQL Browser Service, which allows for connections to named instances of SQL Server that use dynamic ports. Using SQL Browser Service allows us to connect to a database without knowing what port each named instance is using. We will modify the default port for an instance, and disable the SQL Browser Service so that the only way for an attacker to find the TCP port used by our SQL instances is to perform port scanning (which is easier to detect).

There is a TechNet post that talks about a similar solution, *Deploying a SQL Server nonstandard port and alias in Lync Server 2013*, at `http://technet.microsoft.com/en-us/library/dn776290.aspx`. However, if we have more than a single instance on the same SQL Server, it makes sense also to disable the SQL Browser Service. If the service is running, discovering its TCP port will also give information about the ports used by the various instances.

How to do it...

1. On the machine that hosts our SQL server, open **SQL Server Configuration Manager** and go to **SQL Server Network Configuration**. Select **Protocol for "name of our SQL instance"**.
2. Right-click on **TCP/IP** and select **Properties**:

Chapter 1

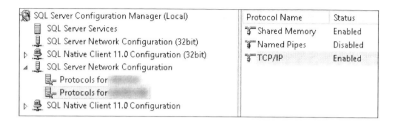

3. Click on the **IP Addresses** tab, select the various IP addresses available for our SQL server, and set **TCP Dynamic Ports** as empty. Set **TCP Port** to the port value we want to use:

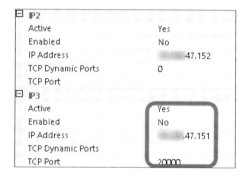

4. Go to the start screen and type Services. Open the services MMC and right-click on the **SQL Server Browser** service. Select **Properties**, and from the drop-down menu, set **Startup Type** as **Disabled**:

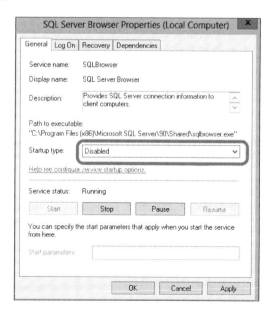

5. On one of the Lync Servers that require access to the database, go to the start screen and type `cliconfg.exe`.

6. Click on the **Alias** tab and select **Add**, as shown in the following screenshot:

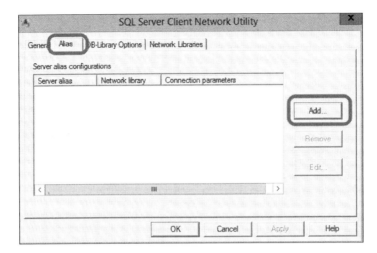

7. In the **Server alias** field, type a name for the SQL alias. In **Network libraries**, select TCP/IP. The **Connection parameter** option is the **Fully Qualified Domain Name** (**FQDN**) of the SQL server\name of the instance. If we have configured a static port for SQL, deselect **Dynamically determine port** and add the port number, as shown in the following screenshot:

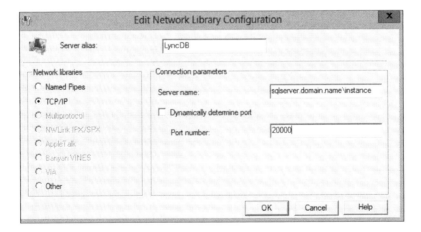

8. Now, go to the start screen and type `regedit`. Navigate to `HKEY_LOCAL_MACHINE\SOFTWARE\Microsoft\MSSQLServer\Client\ConnectTo` and then right-click and select **Export**:

Chapter 1

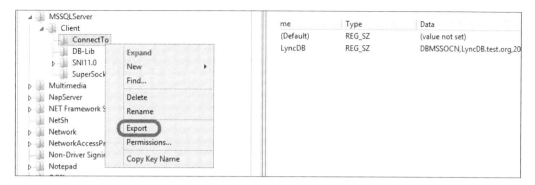

Now, we are able to use the `.reg` file to import the same server alias settings on all the Lync Servers that require a connection to the database.

How it works...

Customizing and limiting the TCP/IP service ports used by SQL server will make it easier to protect the database, especially when we are using a firewall to protect the server. The SQL Server Browser service answers to requests for SQL resources and redirects the caller to the port where SQL server is listening. If this service is disabled, an external attack will be more complex. Aliases will be used in the Lync Topology to connect our deployment to the databases that we have secured.

There's more...

As we mentioned before, there are other ways to protect our database, for example, at the single file level, using a SQL server feature known as **Transparent Data Encryption** (**TDE**). TDE performs real-time encryption and decryption of the data and logfiles. It is supported in Lync Server 2013 as stated in the *Lync Server 2013 supports TDE in SQL Server 2008 or a later version on a backend server* post found at `http://support.microsoft.com/kb/2912342`.

See also

- On the TechNet wiki, there are a couple of interesting posts such as *Database Engine Security Checklist: Enhance the Security of Database Engine Connections* at `http://social.technet.microsoft.com/wiki/contents/articles/1257.database-engine-security-checklist-enhance-the-security-of-database-engine-connections.aspx` and *Database Engine Security Checklist: Database Engine Security Configuration* at `http://social.technet.microsoft.com/wiki/contents/articles/1256.database-engine-security-checklist-database-engine-security-configuration.aspx` full of hints to enhance SQL server security

- The Microsoft site also contains documentation regarding *SQL Server 2008 R2 Security Best Practices* (http://download.microsoft.com/download/1/2/A/12ABE102-4427-4335-B989-5DA579A4D29D/SQL_Server_2008_R2_Security_Best_Practice_Whitepaper.docx) and *SQL Server 2012 Security Best Practices* (http://download.microsoft.com/download/8/F/A/8FABACD7-803E-40FC-ADF8-355E7D218F4C/SQL_Server_2012_Security_Best_Practice_Whitepaper_Apr2012.docx)

Enhancing conferencing security

Conferencing, in Lync Server 2013, has the same default security configuration that we had in Lync Server 2010. Justin Morris, in a post dedicated to Lync Server 2010 conferencing (http://www.justin-morris.net/understanding-conference-security-in-lync-server-2010/), pointed out some of them, including the fact that every Lync user has a meeting URL and a conference ID that never changes by default, and that users invited to a meeting using a PSTN line have the ability to bypass the lobby. If we always use the same conference ID, people outside our company whom we have invited in the past could use it with malicious motivations, or different (unrelated) conferences with different people could merge if our time schedule is not perfect. We will list some steps to improve conferencing security, including a couple of cmdlets from the previously mentioned post.

How to do it...

1. We can start assessing the security configuration of conferencing for our Lync deployment using `Get-CsMeetingConfiguration`.

2. To disable the lobby bypass for PSTN users, use the following cmdlet:

   ```
   Set-CsMeetingConfiguration -PstnCallersBypassLobby $false
   ```

3. To force users to schedule private meetings (with unique IDs), use the following cmdlet:

   ```
   Set-CsMeetingConfiguration -AssignedConferenceTypeByDefault $false -EnableAssignedConferenceType $true
   ```

4. Then, we have to work on the conferencing policies. The default global policy allows almost any of the available conferencing features. I suggest that you clone it using `New-CsClonedPolicy` from Pat Richard (http://www.ehloworld.com/2300). This script is useful to create a new user-conferencing policy called, for example, `DefaultConferencing`, which will have the same parameters we had in the global policy. The cmdlet we will use is the following one:

   ```
   .\New-CsClonedPolicy.ps1 -SourcePolicyName global -TargetPolicyName "DefaultConferencing" -PolicyType ConferencingPolicy
   ```

5. Restrict the global conferencing policy to match the minimal level of service we will grant to every Lync-enabled user.

6. To enhance conferencing security, we can also restrict the number of TCP ports involved in the client communication. Launch the `Get-CsConferencingConfiguration` cmdlet to look at the applied configuration. If `ClientMediaPortRangeEnabled` is `False`, then the clients will use a port between ports `1024` and `65535` when involved in a communication session. We can use the following cmdlet to force port range (default will be starting from TCP port `5350` for all the kind of network traffic):

```
Get-CsConferencingConfiguration | Set-CsConferencingConfiguration
-ClientMediaPortRangeEnabled $True
```

How it works...

As we said, the cmdlet shown in step 6 will use the same TCP port for all the network traffic involved in the clients' communication. We can refer to *Chapter 11, Controlling Your Network – A Quick Drill into QoS and CAC* to also implement **Quality of Service** (**QoS**) and **Call Admission Control** (**CAC**).

See also

I suggest that you also read a second post from Justin Morris, which is related to the unique conference IDs, *Why You Should and Shouldn't Configure Unique Conference IDs in Lync* (http://www.justin-morris.net/why-you-should-and-shouldnt-configure-unique-conference-ids-in-lync/#sthash.W9h9mCzn.dpuf). This post examines the pros and cons of the solution outlined in the previous document in depth.

Managing certificates for the authentication of desk-phones

Lync Phone Edition uses digital certificates during the log-on phase to initialize the connection with the Lync Server. The next step for the logon with a desk phone will be to verify the user with a password (or PIN) authentication. In our deployment, we usually have certificates that come from more than one **Certification Authority** (**CA**), including internal and third-party CAs. The scenario becomes increasingly complex if we are going to use a CA for Exchange Server that is different from the one we used for our Lync Server certificate. Lync Phone Edition usually has a limited number of embedded root CA certificates that are already trusted. The list is available in the Trusted Authorities Cache paragraph in the TechNet post, *Certificates for Lync Phone Edition* found at http://technet.microsoft.com/en-us/library/gg398270(v=ocs.14).aspx. We have to work with the Certificate Provisioning Service to deploy additional root CA certificates.

Getting ready

We will use a Lync Standard Edition certificate issued from an internal CA as an example. Kevin Peters did a good job explaining the process for Lync Server 2010 at http://ocsguy.com/2012/05/19/lync-phone-edition-connection-to-microsoft-exchange-is-unavailable/. We have to follow the same steps for Lync Server 2013.

How to do it...

1. Open **MMC**, navigate to **Add or Remove Snap-ins**, and add **Certificates**.
2. Select **Computer Account** and go to **Local Computer**, and then click on **Finish**.
3. Navigate to **Certificates (Local Computer) | Personal | Certificates** and select the server certificate.
4. Go to **Certification Path** and select the root CA, as shown in the following screenshot:

5. Select the **Details** tab and go to **Thumbprint**. Copy the thumbprint value (*Ctrl + C*).
6. Paste the value in the notepad, remove the empty spaces, and copy the new value.
7. Launch the **Lync Management Shell** and type `$cert = New-CsWebTrustedCACertificate -Thumbprint "⊠Thumbprint" -CAStore TrustedRootCA`.

   The thumbprint is the value we copied at step 6. For example, consider the following:

   `$cert = New-CsWebTrustedCACertificate -Thumbprint "7a06f5b75287f17d4596118418b77004b4cd4d92" -CAStore TrustedRootCA`

8. Type the following cmdlet:

   `Set-CsWebServiceConfiguration -TrustedCACerts @{Add=$cert}`

9. Use the following cmdlet to verify that the thumbprint has been added to the `TrustedCACerts` parameter:

   `Get-CSWebServiceConfiguration`

10. Repeat the preceding steps for all the intermediate and root CA required in our deployment.

How it works...

The desk-phones configuration works on the parameters received from the **Dynamic Host Configuration Protocol** (**DHCP**) server. In particular, options 43 and 120 define the path to the Certificate Provisioning Service on the Lync Server. The steps in this recipe are also required to enable the desk phone to open the necessary URI.

> For a deep dive into desk-phone configuration, a blog post from Jeff Schertz is definitely a must read. The *Configuring Lync Server for Phone Edition Devices* post found at `http://blog.schertz.name/2010/12/configuring-lync-server-for-phone-edition-devices/` contains a complete overview of the parameters required to make Lync desk phones work in a smooth manner.

Deploying a secure Lync Edge

Lync Edge is a role that makes Lync services available to external users and companies in a secure manner. A Lync Edge Server is not part of the internal domain, and in general, it is deployed in a **Demilitarized Zone** (**DMZ**) network with a series of limitations (for example, usually, the name resolution of the Lync Servers in the internal network is limited to local hosts file). The previously mentioned restrictions and other security features, however, are not enough to defend the server from some kind of Internet threats including **Distributed Denial-of-Service** (**DDoS**) and brute-force attacks. An extremely powerful instrument we have at our disposal to protect our Lync deployment is the **Microsoft SIP Processing Language** (**MSPL**). MSPL is a scripting language used specifically to filter and route SIP messages. Chris Norman, for example, has shared a really interesting MSPL script to block presence indicators with federated partner at `http://voipnorm.blogspot.it/2013/01/mspl-script-blocking-federated-presence.html`. The idea behind the script is to keep features we need running (such as IM), limiting the information we show to external users. The steps we will see (required to apply the previously mentioned solution) are the same every time we use an MSPL script.

How to do it...

1. Connect to a Lync Front End Server.

2. Open the Lync Server Management Shell and launch the following cmdlet to get a list of the existing server applications:

   ```
   Get-CsServerApplication
   ```

3. Now, we can install the script (`BlockFederatedPresence.am`).

   ```
   New-CsServerApplication -Identity "EdgeServe:cheshirecat.
   absoluteuc.corp/Simple" -Uri "http://sip.absoluteuc.org/
   BlockFederatedPresence" -ScriptName "C:\BlockFederatedPresence.
   am"-Enabled $true -Critical $false
   ```

   The explanation of the preceding code is as follows:

   - The `Identity` parameter indicates the Edge Server using the FQDN of the server. In our scenario, it is the Edge Server `cheshirecat.absoluteuc.corp`.
   - Uri is a value we define in the `appUri` parameter of the script (I customized the one in Chris Norman's script found at `http://sip.yourdomain.com/BlockFederatedPresence`).
   - ScriptName indicates the path to the `.am` file (in our scenario, `C:\BlockFederatedPresence.am`).
   - Enabled `$true` or `$false` is a switch to activate or deactivate the script.
   - Critical `$true` or `$false` is a switch to select whether Lync Server can start when our application does not start for any reason.

4. We have to restart the Lync services (`Stop-CsWindowsService | Start-CsWindowsService`).

See Also

Rui Maximo wrote a post for Lync 2010 Edge Servers where he talks about DDOS to the Lync Edge Servers and points out how to use MSPL to handle this kind of threat Lync Server 2010: *Security at the Edge* at `http://technet.microsoft.com/en-us/magazine/hh219285.aspx`, *Code4Lync* at `http://mohamedasakr.wordpress.com/`, and *Lync Development* at `http://blog.greenl.ee/` are two blogs, from Mohamed Sakr and Michael Greenlee, that have a lot of interesting posts dedicated to MSPL.

Chapter 1

Applying ethical walls for federation security

The **Instant Messaging** (**IM**) feature of Lync and the capability to communicate with external companies using Federation expose our Lync deployment to a series of threats and compliance issues. In many companies, it is required (for compliance or organizational motivation) that not all people inside all departments are able to use IM to talk to each other. Also, if our company uses federation, there is an increased risk of improper or unprofessional use of Lync features that involve external connections. Talking about IM, the first tool we have for the purpose of mitigating the previously mentioned risks is the Intelligent IM Filter that enables us to block files and URLs in IM conversations. We will see some examples to create additional controls on features.

How to do it...

1. To disable inline images (pictures added to IM conversations), we can use the two filters proposed by Graham Cropley (http://www.lyncexch.co.uk/love-lync-26th-sept-2013/).

2. To block .jpg files, use the following:

   ```
   Set-CsFileTransferFilterConfiguration -Identity <Site / Global> -Extensions @{Add=".jpg"}
   ```

3. To disable Rich Text Format, use the following:

   ```
   Set-CsClientPolicy -Identity <policy> -DisableRTFIm $true
   ```

4. Before we go to the next step, it is useful to know a list of file extensions that are prohibited from our policies. The cmdlet to use is:

   ```
   Get-CsFileTransferFilterConfiguration -Identity global | Select-Object –ExpandProperty Extensions
   ```

5. According to Chris Norman (http://voipnorm.blogspot.it/p/about.html), we have three filters that we can use to enhance the security for our Lync Edge Server (we can pair them with the steps in the *Deploying a secure Lync Edge* recipe). The following scripts could be used respectively to block URL style links, to block the transfer of files with common extensions, or to block the transfer of all files:

   ```
   New-CsImFilterconfiguration -Identity EdgePoolIdentity -BlockFileExtension $True -Action Block -Enabled $True
   New-CsFileTransferFilterConfiguration -Identity EdgePoolIdentity -BlockFileExtension $True -Action Block -Enabled $True
   New-CsFileTransferFilterConfiguration -Identity EdgePoolIdentity -BlockFileExtension $True -Action BlockAll -Enabled $True
   ```

There's more...

By default, Lync tries to establish what kind of relationships we have with a specific person, and grants different rights to that person. This is called **Privacy Relationships**, and it gives control to every single user to establish what kind of information they want to share with a specific contact, varying from a complete opening with **Friends and Family** to a total exclusion for **Blocked Contacts**. In the following screenshot, we are able to see the **Change Privacy Relationship** menu inside the Lync desktop client:

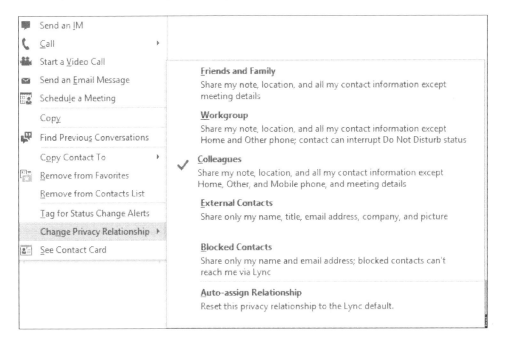

Downloading the example code

You can download the example code files from your account at http://www.packtpub.com for all the Packt Publishing books you have purchased. If you purchased this book elsewhere, you can visit http://www.packtpub.com/support and register to have the files e-mailed directly to you.

On the Microsoft Office pages (`http://office.microsoft.com/en-us/communicator-help/control-access-to-your-presence-information-HA101850361.aspx`), there is a handy table that shows the information available to contacts in the different privacy relationship groups. We can see a copy of it in the next screenshot:

PRESENCE INFORMATION	EXTERNAL CONTACTS	COLLEAGUES	WORKGROUP	FRIENDS AND FAMILY
Presence Status	•	•	•	•
Display Name	•	•	•	•
Email Address	•	•	•	•
Title *	•	•	•	•
Work Phone *		•	•	•
Mobile Phone *			•	•
Home Phone *				•
Other Phone				•
Company *	•	•	•	•
Office *	•	•	•	•
SharePoint Site *		•	•	•
Meeting Location #			•	
Meeting Subject #			•	
Free Busy		•	•	•
Working Hours		•	•	•
Location #		•	•	•
Notes (Out-of-Office Note)		•	•	•
Notes (Personal)		•	•	•
Last Active		•	•	
Personal Photo Web Address	•	•	•	•

This kind of solution manages the problem at the user level and leaves out many scenarios. For example, if we need to apply more granular control (like issuing a warning to users that are typing sensitive data such as credit card numbers), we have no out-of-the-box tool. Segregating different groups of our users (or having a stricter control over companies that are going to federate with us) is also a problem. In the previously mentioned scenarios, what we need is a so-called ethical wall, usually adopting a solution developed by a third-party provider.

Using Application Request Routing to configure a reverse proxy for Lync Server 2013

A reverse proxy (a service that retrieves information on behalf of a client from servers usually located in a corporate network) is a required part of any Lync Server 2013 deployment that we want to make available to external users (as shown at http://technet.microsoft.com/en-us/library/gg398069.aspx). There are many reverse proxy solutions offered by vendors and a couple of free solutions available as part of the Windows OS. The first one is the **Web Application Proxy** (**WAP**) that is part of the Remote Access Server role in Windows Server 2012 R2. However, WAP is a relatively new technology, and there have been some issues related to using it as a reverse proxy solution for Lync Server 2013. The other solution is the IIS **Application Request Routing** (**ARR**) that enables a Windows server to act as a reverse proxy. We will talk about the setup required for the latter scenario.

Getting ready

The latest available version of IIS ARR is 3.0. A prerequirement for ARR 3.0 is to have IIS enabled on the server (we can use the default settings). ARR is available using the **Microsoft Web Platform Installer** (**MWPI**) at http://www.microsoft.com/web/gallery/install.aspx?appid=ARRv3_0. There is also a standalone version (ARR 3.0 standalone package) at http://www.microsoft.com/en-us/download/details.aspx?id=40813, which does not require the Web Platform Installer and could be a good solution for a server with a high level of segregation. We will use the first option (with MWPI). The FQDN of the server we are going to use is dormouse, the Lync 2013 Standard Edition Front End server is madhatter.absoluteuc.corp, and the AD DS server (that includes a certification authority) is alice.absoluteuc.corp. The frontend has a public name madhatter.absoluteuc.org, as shown in the following screenshot:

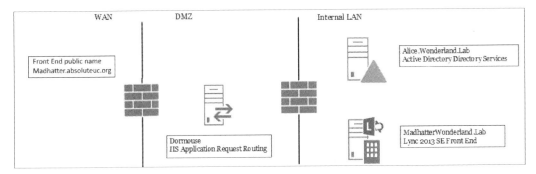

The configuration of a reverse proxy for Lync Server 2013 requires a digital certificate (usually issued from a public CA) that must include at least the Lync Pool FQDN and the Lync simple URLs for Lync (meet and dial-in). The admin simple URL is optional, and for internal use only (http://technet.microsoft.com/en-us/library/gg425874.aspx).

 The reverse proxy in Lync can also be used as a proxy also the simple URLs for Lync Web External (lyncwebext), Mobility (Lyncdiscover), Office Online external URL (which could be unique or shared with lyncwebext). If we enable it, we will also have the simple URL for webscheduler.

Some reference material on this topic is available in the *See also* section of this recipe. A reverse proxy is usually deployed in a **Demilitarized Zone** (**DMZ**) network and will require two different **network interfaces** (**NICs**), one behind the Internet-facing firewall (external NIC) and one behind the backend firewall, used to connect to our internal Lync deployment (internal NIC). The reverse proxy will be reachable from the Internet using a **Network Address Translation** (**NAT**), a process performed by the firewall that converts an internal network address into a different address that is public on the Internet.

How to do it...

1. Our reverse proxy will be located in a DMZ (that is a best practice). Usually, in a similar scenario, we have no access to a DNS server that is able to resolve the FQDNs of the internal domain (in our situation, wonderland.lab), so we have to populate the local HOSTS file on our server.

2. Prepare the SSL digital certificate, starting from a CSR generated using IIS (http://technet.microsoft.com/en-us/library/cc732906(v=ws.10).aspx) or a third-party utility like the DigiCert Certificate Utility for Windows (https://www.digicert.com/util/) and follow the steps required by the selected CA.

3. Open the **Internet Information Services** (**IIS**) Manager and go to the server home screen. Select **Server Certificates** and select **Import...** to import the SSL certificate (in our example, it will be named LyncCert).

4. Open the **Default Web Site** page, and select **Bindings...**.

5. Select **Add**, set **Type HTTPS**, and select the SSL certificate as shown in the following screenshot:

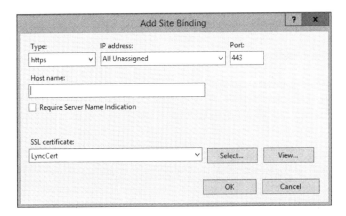

6. Launch the installer and select **Application Request Routing 3.0**, as shown in the following screenshot. The configuration will go on automatically.

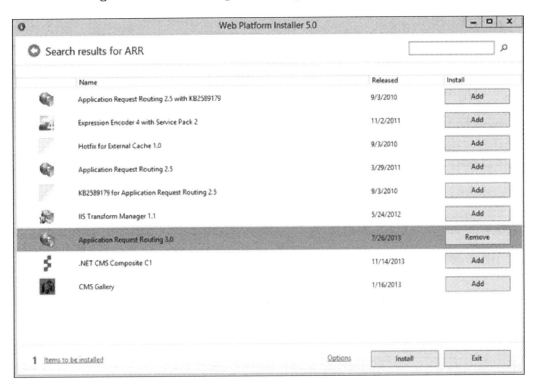

Chapter 1

7. Open the start screen, select **Internet Information Server (IIS) Manager**, and go to the **Server Farms** screen. Right-click and select **Create Server Farm**, as we can see in the following screenshot:

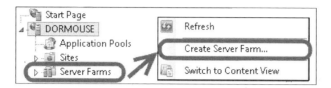

8. Type the FQDN of the service we are going to publish (for example, `madhatter.absoluteuc.org`). Leave the online flag selected. Click on **Next**.

9. Before we add the server, select the **Advanced Settings...** tab. On the **ApplicationRequestRouting** parameters, change the **HTTP** and **HTTPS** ports (for Lync Server 2013, they are `8080` and `4443`, if we haven't customized them in the topology design). In the server name field, type the name of our internal server (or load balancer) and click on **Add**. The configuration is shown in the screen capture. The steps are shown in the following screenshot:

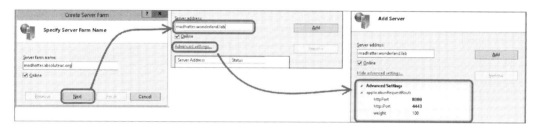

10. Select **Finish**. A notification like the one in the following screenshot will be shown. Select **Yes**.

11. Now, under the **Server Farms** menu, we will have **lyncdiscover.absoluteuc.org**. We have to select it and modify some of the options shown in the following screenshot:

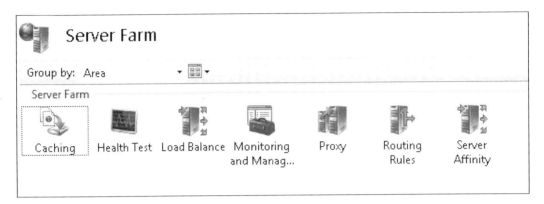

12. Go to the **Caching** options. Remove the flag from **Enable disk cache**. Select **Apply**.
13. Go to the **Routing Rules** options. Remove the flag from **Enable SSL Offloading**. Select **Apply**.
14. Go to the server's home screen and select **URL rewrite**. The previous wizard created two rewrite rules, one with HTTPS/SSL and one with plain HTTP. Remove the one we are not going to use and edit the other one to match our needs. In our example, we are going to publish **madhatter.absoluteuc.org** using HTTPS, so we will remove the HTTP rule and edit the SSL rewrite rule, as shown in the following screenshot:

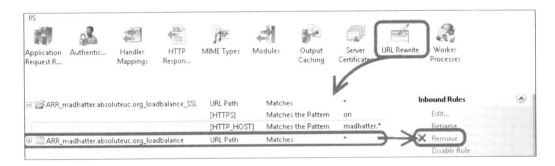

15. The conditions to use are as follows:

    input= {HTTPS}, Type=Matches the Pattern, Pattern=on input={HTTP_HOST}, type=Matches the pattern, Pattern=madhatter.*

16. We also have a **Test Pattern** option to match the URL we are going to use with the rules. When we have done, we have to select **Apply**. The screen is shown in the following screenshot:

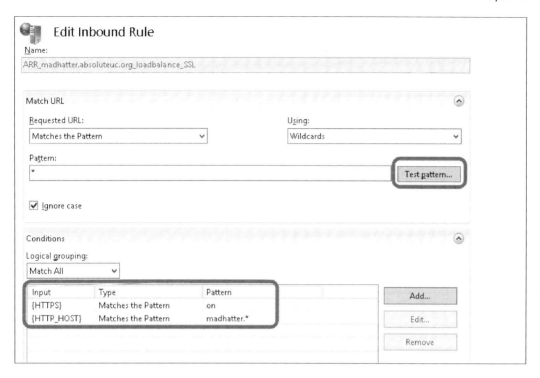

17. The previously mentioned steps have to be repeated for every URL we plan to publish.

> To avoid the "Your server configuration has changed. Please restart Lync" message on Apple mobile devices, it is recommended that you set the Proxy Time-out to 960 seconds (it is a parameter related to every Server Farm we have defined), as we can see in the next screenshot:
>
>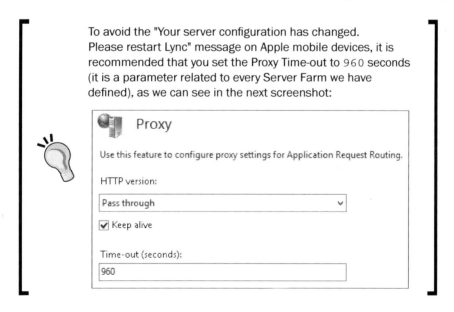

How it works...

The reverse proxy will receive requests on standard ports (TCP 80 for HTTP and TCP 443 for HTTPS) and redirect them to ports 8080 and 4443 of our Lync Server deployment. The "public" Lync 2013 site that is listening on these ports has rewrite rules to handle all the SIP domains we have deployed. For more information, see the *Lync 2013 Internal and External Web Sites* post at `http://tsoorad.blogspot.it/2013/04/lync-2013-internal-and-external-web.html` by John Weber.

See also

To plan Lync certificates, the following are some useful resources:

- *TechNet post Request and Configure a Certificate for Your Reverse HTTP Proxy* at `http://technet.microsoft.com/en-us/library/gg429704.aspx`
- *Lync 2013 Detailed Design Calculator* at `http://gallery.technet.microsoft.com/lync/Lync-2013-Standard-Edition-324bf0f1`
- *Microsoft Lync Server 2013 Internal Certificate Planning and Deployment* at `http://lyncuc.blogspot.de/2014/04/internal-certificate-deployment-in-lync.html`

2
Lync 2013 Authentication

In this chapter, we will cover the following topics:

- Introduction
- Configuring passive authentication for Lync
- Enabling two-factor authentication
- Adding the app password for mobile clients
- Authenticating with online services using DirSync
- Managing Windows Azure Directory for Lync Online
- Configuring server-to-server authentication
- Troubleshooting with client authentication logging

Introduction

Lync Server identity validation mechanisms are able to manage different kinds of personas, including authentication for a device, server, and user. The previously mentioned identities can require access to Lync in an on-premises datacenter, in a public Cloud, or in a hybrid Cloud scenario. The shift from Lync deployments that span over a corporate network to more complex setups required identity verification instruments to be adequate. Windows 2012, Windows 2012 R2, Azure, and Office 365 support a number of solutions (such as DirSync and identity federation) that add flexibility to our design.

Lync Server 2013 is able to accept both Cloud identities (stored in the Cloud) and federated identities (stored and managed on-premises). Talking about Lync Server 2013 on-premises, we have support for the following different user authentication protocols:

- **Kerberos v5** and **NT LAN Manager** (**NTLM**) are reserved for users with Active Directory credentials (Kerberos for users on the internal network and NTLM when the Lync users connect from an external network).

- While an older version of Lync for mobile supported only NTLM for authentication, today's mobile apps/clients use a more secure approach. The initial authentication can be NTLM or passive and the authentication process the client obtains a Lync client certificate from the Lync Certificate Provisioning Service on the Front End server. The certificate on the client will be used for reauthentication, reducing the exposure of Active Directory credentials.

- Anonymous users (with no Active Directory credentials) authenticate via the Digest protocol. Microsoft has also added support for passive authentication in Lync 2013.

- Kerberos is available to external users as well if they connect through a VPN. Using a VPN, however, has some limitations, including a high overhead that could impact audio and video quality.

- Passive authentication (added with **Microsoft Lync 2013 for Mobile release 5.2**) uses form-based authentication built on **Active Directory Federation Services** (**AD FS**). Passive authentication is really interesting from a security point of view because the password is no longer entered on the mobile client. The user signs in by typing only their Lync address and then gets redirected to AD FS for authentication.

Adding services from the Cloud (Azure and Office) 365, we are able to add security and authentication options:

- Same Sign On and Single Sign On, based on DirSync and AD FS. Both the scenarios allow us to insert our on-premises domain account information once and also use it in the cloud.

- Two-factor authentication (like smart cards) adds a layer of security. It is available only for Lync Online users but can be implemented for on-premises using third-party integration (for example, see this post *Lync Passive Authentication with two-factor authentication – Part II* at http://techmikal.com/2014/02/27/lync-passive-authentication-with-two-factor-authentication-part-ii/).

- App passwords and an additional security option for Office 365 are used to replace the account password.

We will see passive authentication, Same Sign-On and Single Sign-On, two factors authentication, and app passwords in specific paragraphs of this chapter. Before we start with the various recipes, we should take a look at the following schema that outlines the lab environment we have used:

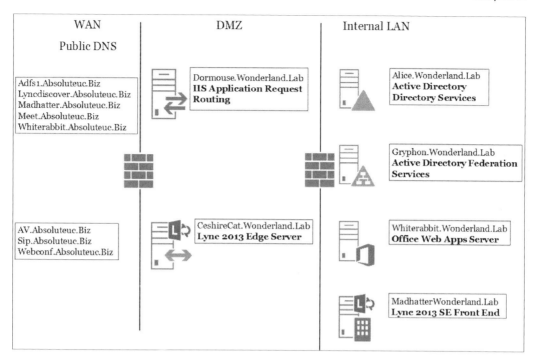

Configuring passive authentication for Lync

Passive authentication requires establishing trust between the AD FS Server and the Lync Server (because user credentials will be forwarded by the latter service). Lync mobility policies require customization too. We will see the steps to configure passive authentication now.

Getting ready

Based on the schema of our lab environment, we will use **madhatter.wonderland.lab** as the internal FQDN of the Lync 2013 Standard Edition (SE) Server. The **madhatter.absoluteuc.biz** is the public name for the pool, while **gryphon.wonderland.lab** is the internal FQDN of the AD FS server, and **adfs1.absoluteuc.biz** is the public name for the same service.

If we have no AD FS server available, the first step will be to add the role from **Server Manager** and configure it (for example, we can follow the steps outlined in the TechNet post *How To Install ADFS 2012 R2 For Office 365* at `http://blogs.technet.com/b/rmilne/archive/2014/04/28/how-to-install-adfs-2012-r2-for-office-365.aspx`).

Lync 2013 Authentication

How to do it...

1. We have to configure trust relations between the AD FS server and the Lync Server so that Lync will accept the account information sent from the AD FS server on behalf of the user. This is done by creating a relaying party trust and appropriate Issuance Authorization and Transform rules (more details about AD FS relying party trust are in the *How it works...* section of this recipe).

2. AD FS trust relationships are based on the server names stored in the SSL certificates. It is required to have two separate trusts, one for the public name of our Lync pool and one for the internal one. We will call them `Lync-Ext-PassiveAuth` (based on the external name) and `Lync-PassiveAuth` (based on the internal name).

    > We is not allowed to create two trust relations that point to the same IP (in this scenario, the one used by the Lync 2013 SE Server), and we have to use the reverse proxy for both Lync mobile clients that connect from the external network and from the internal network. We can split the trusts, one on the external interface of IIS (Lync-Ext-PassiveAuth) and the other hair pinning on the internal interface of IIS (Lync-PassiveAuth).

3. In the following steps, we will define the transform rules and trusts. We have to define the authorizing claims requesters (Authorization Rules) to control which users have access to the relying party:

    ```
    $IssuanceAuthorizationRules = '@RuleTemplate = "AllowAllAuthzRule"
    => issue(Type = "http://schemas.microsoft.com/authorization/
    claims/permit", Value = "true");'
    ```

4. The next step is to define Issuance Transform Rules to issue claims to the relaying party:

    ```
    $IssuanceTransformRules = '@RuleTemplate = "PassThroughClaims"
    @RuleName = "Sid" c:[Type == "http://schemas.microsoft.com/
    ws/2008/06/identity/claims/primarysid"]=> issue(claim = c);'
    ```

5. Now, we are able to define the Relying Party Trust (Lync-Ext-PassiveAuth) between the AD FS server and the internal name of the Lync Server:

    ```
    Add-ADFSRelyingPartyTrust -Name Lync-PassiveAuth
    -MetadataURL https://madhatter.wonderland.lab/passiveauth/
    federationmetadata/2007-06/federationmetadata.xml
    ```

6. The next step is to apply the previously defined Authorization and Issuance Transform Rules to the Lync-PassiveAuth Relying Party Trust:

    ```
    Set-ADFSRelyingPartyTrust -TargetName Lync-PassiveAuth
    -IssuanceAuthorizationRules $IssuanceAuthorizationRules
    Set-ADFSRelyingPartyTrust -TargetName Lync-PassiveAuth
    -IssuanceTransformRules $IssuanceTransformRules
    ```

7. The steps that we performed for the Lync-PassiveAuth trust will be repeated for the **Lync-Ext-PassiveAuth** (that is based on the public name of the Lync Server):

   ```
   Add-ADFSRelyingPartyTrust -Name Lync-Ext-PassiveAuth
   -MetadataURL https://madhatter.absoluteuc.biz/passiveauth/
   federationmetadata/2007-06/federationmetadata.xml
   Set-ADFSRelyingPartyTrust -TargetName Lync-Ext-PassiveAuth
   -IssuanceAuthorizationRules $IssuanceAuthorizationRules
   Set-ADFSRelyingPartyTrust -TargetName Lync-Ext-PassiveAuth
   -IssuanceTransformRules $IssuanceTransformRules
   ```

8. The result is seen in the following screenshot (taken from the **ADFS Management Console**):

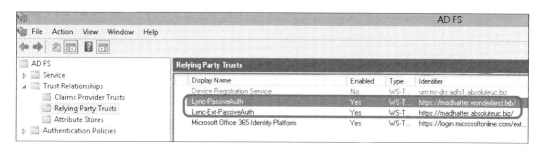

9. It is required to disable other authentication methods on Lync 2013 (we mentioned them in the introduction of this chapter) because the server uses them before using the passive authentication. We have to disable Kerberos and NTLM on three Lync services: **Lync Autodiscover WebServer**, **Front End WebServer** and **Front End Registrar**. The first two will use certificate and passive authentication, while the Front End Registrar will only use certificates.

10. We are going to define a new web service:

    ```
    New-CsWebServiceConfiguration -Identity webserver:madhatter.
    wonderland.lab -WsFedPassiveMetadataUri https://adfs1.absoluteuc.
    biz/federationmetadata/2007-06/federationmetadata.xml
    ```

11. On the new web service, we will enable certificate and passive authentication and disable Kerberos and NTLM:

    ```
    Set-CsWebServiceConfiguration -Identity webserver:madhatter.
    wonderland.lab -UseWsFedPassiveAuth $true -UseWindowsAuth none -
    UseCertificateAuth $true
    ```

12. We will define a new registrar configuration with Kerberos and NTLM disabled:

    ```
    New-CsProxyConfiguration -Identity registrar:madhatter.
    wonderland.lab -UseKerberosForClientToProxyAuth $false
    -UseNtlmForClientToProxyAuth $false
    ```

Passive authentication, when enabled, works at the Lync pool level. This implies that if we have clients that are not compatible with passive authentication, or if we want a subset of our users to connect without passive authentication, we should have a second registrar.

13. By looking at the **Registrar** and **Web Service** tabs, we can verify the result of the previously mentioned configuration from **Lync Server Management Shell** or inside **Lync Server Control Panel | Security**:

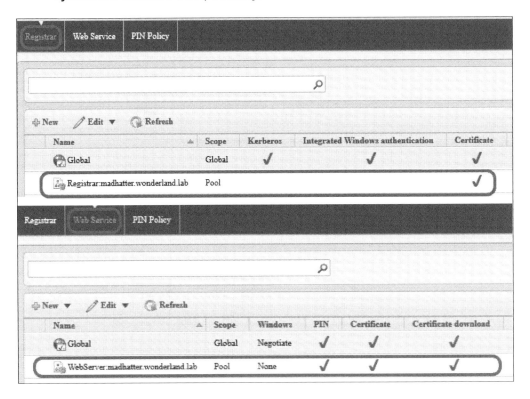

14. Passive authentication for Lync 2013 mobile clients does not support the Exchange server. To enable the passive authentication feature for users, we will set **AllowExchangeConnectivity** to $false in **CsMobilityPolicy** to suppress the error messages related to the **Exchange Web Services** (**EWS**). In our example, we will assign it to the **wonderland\fab** user in the wonderland.lab domain:

```
New-CsMobilityPolicy -Identity PassiveAuthUsers
-AllowExchangeConnectivity $false
Grant-CsMobilityPolicy -PolicyName PassiveAuthUsers -Identity
wonderland\fab
```

15. It is now possible to test passive authentication from a mobile client, like the one for iPad. The logon phase should redirect us to an AD FS form-based authentication page, like the one in the following screenshot:

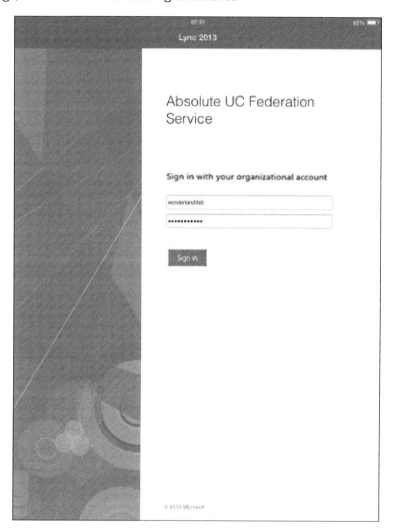

How it works...

To explain how the AD FS Relying Party Trust works, we have to introduce some of the following basic concepts:

- **Claim**: This is information about a user from a trusted source.
- **Trusted source**: This confirms that the content of the claim is true. A trusted source has verified the information with an authentication method.

- **Claims provider**: This is the source of the claim (in our scenario, Active Directory).
- **Relying party**: This is the destination for the claims (in our scenario, Lync Server).
- **Relying party trust**: This is the trust between the AD FS server and the relying party.

Based on a set of rules (**Acceptance Transform Rules**), the claims from the claim provider will be filtered or passed to the Relying party trust. AD FS controls which users have access to the relying party based on the **Issuance Authorization Rules**. AD FS issues the claims to the relaying party by using the **Issuance Transform Rules**. In the schema shown in the following screenshot, we have a high-level overview of the mechanism inside the AD FS server that converts the incoming claims from the **Claims Provider Trust** and transmits them over the **Relying party trust**:

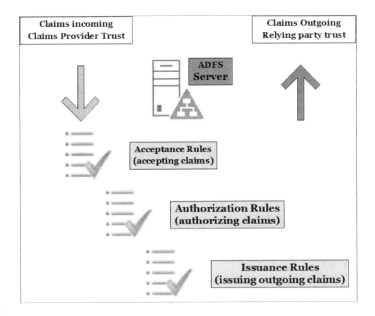

 Right now, Lync clients for Android and Mac are not able to sign in using passive authentication (see *Sign-in, Push Notifications, and General Features* in the *Mobile client comparison tables for Lync Server 2013* TechNet post at http://technet.microsoft.com/en-us/library/hh691004.aspx). Clients with no support for passive authentication will hang in the attempt to authenticate with Lync 2013.

There's more...

How Lync passive authentication works is explained in two blog posts: Jens Trier Rasmussen's TechNet blog *Microsoft Lync 2013 for Mobile and Passive Authentication* at `http://bit.ly/1ujMcSK` and on the TechNet site *Lync Server 2013 Certificate Authentication and Passive Authentication support for Lync 2013 Mobile applications* at `http://bit.ly/1rzopKJ`.

In a recent post on my blog *Lync Passive Authentication: Getting Your Hands Dirty* at `http://bit.ly/WywjJi`, I tried the various clients and tested their behavior with passive authentication.

Enabling two-factor authentication

In 2013, the July cumulative update for Lync Server 2013 added two-factor authentication for the Lync 2013 client. As we mentioned in the introduction, the feature is available for Lync Online users, while on-premises deployment requires you to add a third-party solution. Two-factor authentication is based on the "something you know, something you have" principle, meaning that knowing a user's password without having an additional security item (a smart card) will not give access to Lync. An additional method could be based on "something you have" (a trusted device). We will see how to configure Office 365 (Lync Online) support for **Multi-Factor Authentication** (**MFA**). Office 365 MFA is a scaled-down version of Azure Multi-Factor Authentication service available to support both Azure and on-premises deployments. We will talk about it in the *There's more...* section.

How to do it...

1. Log on to the Office 365 portal and go to **Users and Groups** | **Active users** and select **Set Up** to the right of **Set Multi-factor Authentication Requirements**, as we can see in the next screenshot:

2. Select the user that will use the Multi-Factor Authentication and select **Enable**.

Lync 2013 Authentication

3. We will receive an **About enabling multi-factor auth** screen that requires a confirmation. Select the **Enable Multi-Factor Auth** button, as shown in the following screenshot:

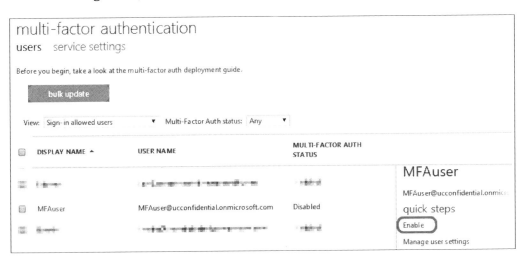

4. Select **Close** in the **Updates Successful** screen.
5. The **Manage User Settings** option enables us to force the user settings related to MFA with the **Require selected users to provide contact methods again** and **Delete all existing app passwords generated by the selected users** features.
6. The next time the user tries to log on to Office 365, they will receive a request to set up this account for additional security verification, as we can see in the following screenshot:

7. There are three available options shown in the following screenshot: **Mobile phone**, **Office phone**, and **Mobile app**. Each one of them requires additional parameters (like a mobile number) to work.

Step 1: Specify the contact method we should use by default

8. The device will be verified with a call or a text message.
9. The next time the user logs on to the Office 365 portal, after the account and password screen, there will be an additional verification based on the selected additional security method.

There's more...

Azure **Multi-Factor Authentication** (**MFA**) is an interesting service, and it is important to say that it can be deployed also in an on-premises environment. Azure MFA includes options for one-time passwords (generated with the MFA app), phone calls, and text messages. Now, we need to perform the following steps:

1. We will need to access the Azure Portal (as explained in the *Managing Windows Azure Directory for Lync Online* recipe).
2. From the directory management portal, we have to launch **Enable multi-factor authentication**, as we can see in the following screenshot:

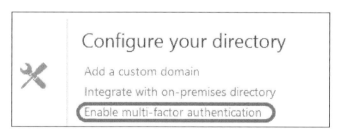

Lync 2013 Authentication

3. The lower bar will show a completed operation notification, like the one shown in the following screenshot:

4. In the lower part of the **USERS** tab, we have the **MANAGE MULTI-FACTOR AUTH** icon:

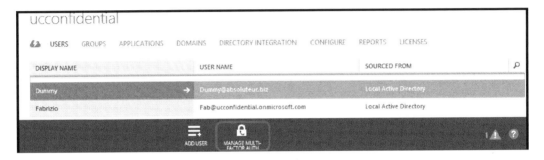

5. As soon as the Multi-Factor Authentication provider for the selected domain is created, we are able to configure the settings for MFA. As we can see in the following screenshot, the feature is richer than the one we saw in Office 365:

Chapter 2

 Azure has two MFA usage models **Per Authentication** and **Per Enabled User**, and both of them have an associated cost. The costs are explained in the Active Directory pricing details page (http://bit.ly/1fspdhz).

Adding the app password for mobile clients

What we mentioned in the *Enabling two-factor authentication* recipe applies also to the mobile apps, but we have to add another mechanism called app password to make it work on these kinds of clients. It is an additional feature required for all the non-browser apps.

How to do it...

1. Lync and other non-browser apps will require an app password, which we are able to generate using the following screen:

Step 3: Apps like Microsoft Office will need new passwords for this account

Your current password will now only work in the browser.

To sign into non-browser apps, you will need a special password called an app password. Examples of these apps are:

- Microsoft Office apps such as Outlook or Lync
- Mobile apps for email

Learn more about app passwords   Bookmark this page

generate app password   i don't use this account with these apps

2. The app password will be shown only once, so remember to save it.
3. From now on, the app password will replace the account password when the user tries to open one of the multifactor-authentication-enabled apps.

Lync 2013 Authentication

How it works...

The user is enabled to generate many app passwords (for example, one for every device) and delete them as required. It is required to access the Office 365 portal with the user, select **Office 365 settings**, go to the **Additional Security Verification** screen, and select the **Update my phone numbers used for account security**, as shown in the following screenshot:

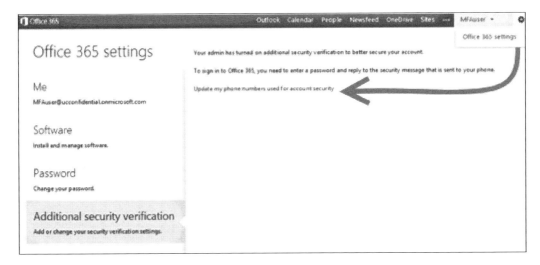

The app password screen contains the required tools to manage this additional security layer.

Authenticating with online services using DirSync

Companies with an on-premises Directory Services deployment often require using the same account information to access both on-premises and Office 365 resources. We have two solutions available to achieve the previously mentioned result.

The first one is named same sign on and it is based on DirSync. It enables the user to log on with the same username and password (if we have also enabled password synchronization) on the legacy data centre and in the cloud. It is easier to deploy this solution because AD FS is not required. The downside is that the user will be required to enter their credentials more than once every time they move from on-premises to the cloud.

Chapter 2

> Sometimes, the password synchronization to the cloud creates security-related concerns. Quoting the TechNet (*Implement Password Synchronization* at `http://technet.microsoft.com/en-us/library/dn246918.aspx`):
>
> *"When synchronizing passwords using the password sync feature, the plain text version of a user's password is (never) exposed. A digest of the Windows Active Directory password hash is used for the transmission between the on-premises AD and Azure Active Directory. The digest of the password hash cannot be used to access resources in the customer's on-premises environment."*

The second solution, Single Sign On, requires both DirSync and AD FS, but enables the user to authenticate once on our domain and then access Office 365 resources with no additional verification required. In the schema shown in the following screenshot, we can see the logic used in the Same Sign On and in the Single Sign On scenarios:

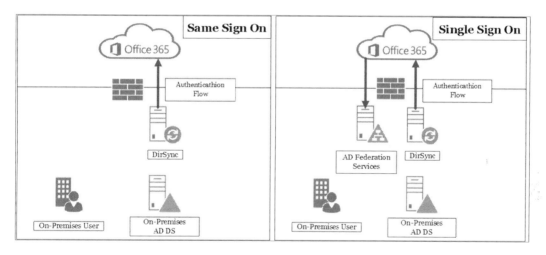

> In the Single Sign On scenario, it is recommended that you deploy two separate servers, one for AD FS and one for DirSync. The motivation for this is that if we collocate the two services, we are not able to use a high availability deployment for AD FS.

Getting ready

We will see the configuration required to use Single Sign On. Regarding the AD FS part of the configuration, the steps required to configure it are the same as we saw previously. Inside your Office 365 subscription portal, there is a **Set up and manage single sign-on** page that is a high-level overview of the required steps. You are able to find it by selecting **Users and Groups | Active users** and opening the **Manage** shortcut for the Single sign-on:

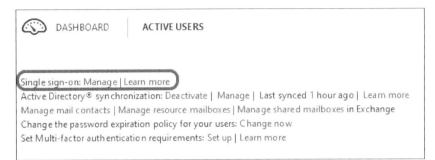

The **Set up** page, however, is only a quick overview, so here we will try to add the missing information.

DirSync installation has some software-specific prerequirements:

- The Windows Server operating system (Windows Server 2008, 2008 R2, 2012, or 2012 R2)
- A server joined to the Active Directory forest that we are going to synchronize
- Microsoft .NET Framework 3.5 SP1 and the Microsoft .NET Framework 4.5.1
- It must run (at least) Windows PowerShell 2.0

If we are going to install DirSync on a domain-joined Windows Server 2012 R2, the only requirement missing is the .NET Framework 3.5, which we are able to add as a Windows feature from Server Manager.

 The DirSync server requires access on TCP port 80 and TCP port 443 (HTTP and HTTPS) on the Internet to communicate with Office 365.

Chapter 2

Usually, the AD DS domain and the public domain have different names. To make it easier for users to use Single Sign On, with a username such as `user@publicdomain` (in our scenario, for example, `dummy@absoluteuc.biz`), we have to enable **User Principal Name** (**UPN**). The feature is available inside **Active Directory Domains and Trusts**, and can be accessed by right-clicking on the root and selecting properties. This screen is shown in the following screenshot:

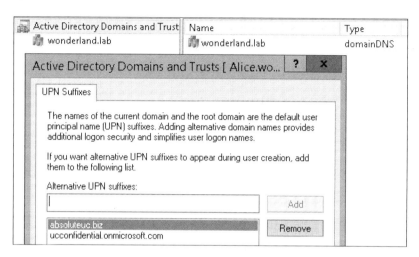

How to do it...

1. Configure an AD FS server accessible from the Internet, with an FQDN published on the public DNS servers and with an SSL certificate issued from a well-known public certificate authority. It is required to create a secure communication channel between the on-premises deployment and Office 365. We have used the `gryphon.wonderland.lab` server, with a public name `adfs1.absoluteuc.biz`. It is exposed by using IIS ARR (`dormhouse.wonderland.lab`).

2. On a server with an Internet connection available, we have to install the following two different components:
   - Windows Azure AD Module for Windows PowerShell
   - The Directory Synchronization tool (DirSync)

    On October 20, 2014, Windows Azure AD Module 32-bit is discontinued. So both DirSync and Windows Azure AD Module have to be in the 64-bit version.

3. To install Windows Azure AD Module, the Microsoft Online Services Sign-In Assistant for IT Professionals RTW (http://www.microsoft.com/en-us/download/details.aspx?id=28177) is required, so we have to install it.

49

Lync 2013 Authentication

4. The next action required to be performed is the installation of the Windows Azure AD Module that is available on Technet at `http://go.microsoft.com/fwlink/p/?linkid=236297` or in the previously mentioned Office 365 page.

   As soon as the setup is complete, you will have the icon shown in the following screenshot:

5. It is required to activate the Active Directory synchronization inside Office 365. This operation is available from the dashboard or inside the Single sign-on setup page introduced in the *Getting ready* section.

6. The setup process for DirSync requires administrative credentials on the server where we are performing the installation, on the local Active Directory, and in our Microsoft Cloud service.

7. Now, we are able to install DirSync. At the end of the installation, we will see a lot of additional software installed, as you can see in the following screenshot. This are required to make DirSync work.

Forefront Identity Manager Synchronization Service	Microsoft Corporation	7/23/2014
Microsoft Office Web Apps Server 2013	Microsoft Corporation	7/15/2014
Microsoft Office Web Apps Server Language Pack 201...	Microsoft Corporation	7/15/2014
Microsoft Online Services Sign-in Assistant	Microsoft Corporation	7/24/2014
Microsoft SQL Server 2008 Setup Support Files	Microsoft Corporation	7/23/2014
Microsoft SQL Server 2012 (64-bit)	Microsoft Corporation	7/23/2014
Microsoft SQL Server 2012 Native Client	Microsoft Corporation	7/23/2014
Microsoft SQL Server 2012 Setup (English)	Microsoft Corporation	7/23/2014
Microsoft SQL Server 2012 Transact-SQL ScriptDom	Microsoft Corporation	7/23/2014
Microsoft Visual C++ 2010 x64 Redistributable - 10.0....	Microsoft Corporation	7/23/2014
Microsoft Visual C++ 2010 x86 Redistributable - 10.0....	Microsoft Corporation	7/23/2014
Microsoft VSS Writer for SQL Server 2012	Microsoft Corporation	7/23/2014
SQL Server Browser for SQL Server 2012	Microsoft Corporation	7/23/2014
Windows Azure Active Directory Module for Window...	Microsoft Corporation	7/24/2014
Windows Azure Active Directory Sync tool	Microsoft Corporation	7/23/2014

8. Back to the users and groups, Active users we should see our domain users in the list, with the **Synced with Active Directory** text on the right, as we can see in the next screenshot:

9. What we have done until now is enough to have Same Sign On enabled. However, we are going to do more and enable Single Sign On.

10. If we have enabled UPNs, we are able to modify user logon names so that they include our domain public name. We can see it in the drop-down menu shown in the following screenshot:

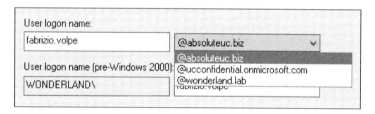

11. Launch the Windows Azure AD Module for Windows PowerShell and use the following statements to register the Office 365 credentials and enable communication between your AD FS and the cloud:

  - `$cred=Get-Credential` (you will be prompted for the username and password of an Office 365 administrator)
  - `Connect-MsolService -Credential $cred`
  - `Set-MsolAdfscontext -Computer "internal FQDN of your AD FS server"`. For example, we have `Set-MsolAdfscontext -Computer Gryphon.wonderland.lab`
  - `Add-MsolDomainToFederated -DomainName "FQDN of our public domain"`

Lync 2013 Authentication

 We need to use the **convert** option for domains that are already defined inside Office 365. Also, if we are going to use multiple domains, we need to add the `SupportMultipleDomain` switch. In our scenario, the cmdlet was `Convert-MsolDomainToFederated -DomainName ucconfidential.onmicrosoft.com -SupportMultipleDomain`.

How it works...

This configuration unifies the account log on. If a user logs on to a domain-joined workstation and tries to log on to Office 365, they will be redirected to our company's authentication page just after typing their username. Then, they will be able to authenticate with the domain user and password, as shown in the following screenshot:

Chapter 2

There's more...

Starting with the June 2013 updates, it is possible to install the **Windows Azure Active Directory** (**WAAD**) **Directory Sync** (**DirSync**) tool on a domain controller. This solution makes sense for companies that want to have as few servers running as possible and prefer not to dedicate a machine to DirSync. However, it is also a solution that Microsoft recommends only in a development environment. The DirSync tool uses a scaled-down version of the Microsoft **Forefront Identity Management** (**FIM**) server and also installs other components such as SQL Express. The overhead associated with the previously mentioned solution is something that we have to consider on a Domain Controller (we should also consider whether we can change some configuration on DirSync, such as the interval between synchronizations, to reduce the overhead).

> In some scenarios, such as companies with a limited number of available servers, it could also be interesting to install AD FS and DirSync on the same server. There is no rule that prohibits such a solution, but we have to consider that the deployment of AD FS (especially if we aim to have a highly available situation) mixed with DirSync increases the complexity of the configuration.

A useful tool to debug Single Sign On is the **Microsoft Remote Connectivity Analyzer** (`https://testconnectivity.microsoft.com/`) **Office 365** and **Office 365 Single Sign-On Test**, as shown in the following screenshot:

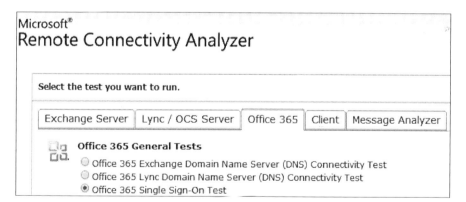

> In *Chapter 7*, *Lync 2013 in a Resource Forest*, we will also talk about Azure **Active Directory Synchronization Services** (**AAD Sync**). This is a recently released solution that could (also) replace DirSync as an account synchronization tool.

Managing Windows Azure Directory for Lync Online

Office 365 uses Azure Active Directory to provide authentication (identities are synchronized with Azure AD to provide authentication). Azure AD is the only security container if you have no on-premises AD DS and no DirSync deployed. The Office 365 portal gives you limited access to the features related to the directory service; mostly, we are restricted to users and group management. Inside the Azure Portal, we have the tools required to work directly on the Azure AD. The following screenshot shows the reports page for Azure Directory, which is really interesting from a security point of view:

REPORT	DESCRIPTION
◢ ANOMALOUS ACTIVITY	
Sign ins from unknown sources	May indicate an attempt to sign in without being traced.
Sign ins after multiple failures	May indicate a successful brute force attack.
Sign ins from multiple geographies	May indicate that multiple users are signing in with the same account.
◢ ERROR REPORTS	
Account provisioning errors	Indicates an impact to users' access to external applications.
◢ PREMIUM REPORTS	
Sign ins from IP addresses with suspicious activity	May indicate a successful sign in after a sustained intrusion attempt.
Sign ins from possibly infected devices	May indicate an attempt to sign in from possibly infected devices.
Irregular sign in activity	May indicate events anomalous to users' sign in patterns.
Users with anomalous sign in activity	Indicates users whose accounts may have been compromised.
Application usage	Provides a usage summary for all SaaS applications integrated with your directory.

We will see how to access your directory through Azure.

Chapter 2

How to do it...

1. Access the Azure Portal using an Internet browser at `https://bitly.com/WJf1Jf`.
2. Log in using the administrative account related to your Office 365 subscription.
3. You will be required to sign up for Windows Azure, so click on **SIGN UP FOR WINDOWS AZURE**, as shown in the following screenshot:

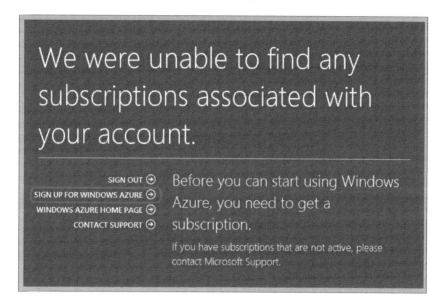

4. After registering for the trial period, you will be enabled with the Office 365 administrator account and password.
5. Go to **all items**, and you will see your domain already enabled, as we can see in the following screenshot:

6. By selecting it, we have access to the features and tools exhibited in the following screenshot, including multifactor authentication and domain joining:

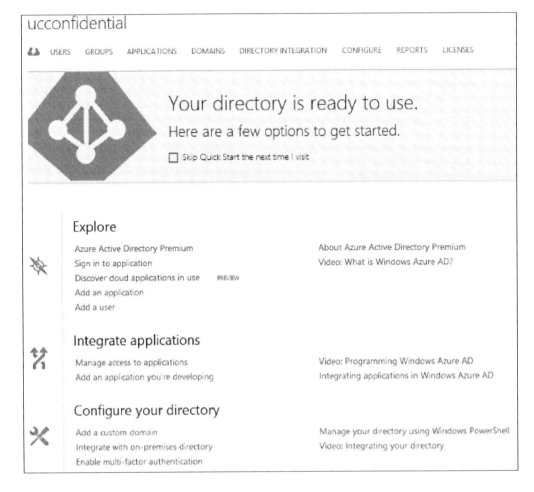

For example, if we have integrated an on-premises directory, we are able to verify and manage it from the **Integrate with on-premises directory** option.

Configuring server-to-server authentication

Lync Server 2013 integrates with many complementary servers (Exchange, SharePoint, and Office Web Apps) to deliver additional features (such as unified contact store and voice mail). Lync is able to use a server-to-server authentication protocol, the **Open Authorization 2.0** (**OAuth 2.0**), to make it easier for the server to act on behalf of the user when required, and to pass a security token from a server to the other one instead of using the user's credentials.

The following are three server-to-server authentication scenarios (note that this kind of authentication is supported only with Exchange 2013, Lync 2013, and SharePoint 2013):

- Server-to-server authentication with Lync Server 2013, Exchange 2013, and SharePoint on-premises
- Server-to-server authentication between two Office 365 components
- Server-to-server authentication in a cross-premises (hybrid) environment

Server-to-server authentication on-premises uses a token server that is built-in Exchange 2013 and Lync 2013. Office 365 servers rely on a third-party token server.

We will see how to configure OAuth for a cross-premises Lync deployment, with our Lync users distributed on-premises and on the cloud. Information is available in this TechNet post *Configuring Microsoft Lync Server 2013 in a cross-premises environment* at `http://bit.ly/1mK1Dts`.

Getting ready

It is necessary to generate an OAuthTokenIssuer certificate that will be applied to the Lync Server. As explained in the *There's more...* section of this recipe, the same certificate will be used on all our Lync Server 2013 servers. The previously mentioned certificate could be a wildcard certificate (in our scenario, `*.absoluteuc.biz`) or a multidomain (SAN) certificate, including the names of all our Lync Front Ends.

It is required to have a DirSync server in our on-premises deployment to manage the Lync Online-enabled users. It can be configured as we saw in the previous sections. It is important to note that the setup includes the Microsoft Online Services Sign-in Assistant and the Microsoft Online Services Module for Windows PowerShell.

> For our scenario, we used a single third-party SAN certificate that includes the name of the DirSync server and the name of the Lync FE server.

How to do it...

1. We have to configure our on-premises Lync Server to accept the Office 365 authorization server. The URI we will use is `https://accounts.accesscontrol.windows.net`, which is the Azure **Active Directory** (**AD**) endpoint for this kind of authentication and authorization.

2. We can use (inside a Lync Management Shell) the statements shown in the previously mentioned TechNet post, replacing the name of the tenant in the first line with our domain name in Office 365 (in our scenario, it will be `$TenantID = "absoluteuc.biz"`). The rest of the script will not require modifications. We will divide the statements in two parts to better understand the process:

```
$TenantID = "absoluteuc.biz"
$sts = Get-CsOAuthServer microsoft.sts -ErrorAction SilentlyContinue
    if ($sts -eq $null)
        {
            New-CsOAuthServer microsoft.sts -MetadataUrl "https://accounts.accesscontrol.windows.net/$TenantId/metadata/json/1"
        }
    else
        {
            if ($sts.MetadataUrl -ne "https://accounts.accesscontrol.windows.net/$TenantId/metadata/json/1")
                {
                    Remove-CsOAuthServer microsoft.sts
                    New-CsOAuthServer microsoft.sts -MetadataUrl "https://accounts.accesscontrol.windows.net/$TenantId/metadata/json/1"
                }
        }
```

> The script to generate the trust relationships between Lync Server 2013 and the authorization server (and vice versa) should run on the DirSync server that has the Windows Azure AD Module for Windows PowerShell installed.

3. The previous statements create a new OAuthServer, Microsoft Security Token Service (Microsoft.sts), which is reachable on the previously mentioned `https://accounts.accesscontrol.windows.net` URL. If another server with the same name already exists, it will be removed:

```
$exch = Get-CsPartnerApplication microsoft.exchange -ErrorAction SilentlyContinue
if ($exch -eq $null)
    {
        New-CsPartnerApplication -Identity microsoft.exchange -ApplicationIdentifier 00000002-0000-0ff1-ce00-000000000000 -ApplicationTrustLevel Full -UseOAuthServer
    }
else
    {
```

```
        if ($exch.ApplicationIdentifier -ne "00000002-0000-0ff1-
ce00-000000000000")
           {
               Remove-CsPartnerApplication microsoft.exchange
               New-CsPartnerApplication -Identity microsoft.exchange
-ApplicationIdentifier 00000002-0000-0ff1-ce00-000000000000
-ApplicationTrustLevel Full -UseOAuthServer
           }
        else
           {
               Set-CsPartnerApplication -Identity microsoft.exchange
-ApplicationTrustLevel Full -UseOAuthServer
           }
    }
Set-CsOAuthConfiguration -ServiceName 00000004-0000-0ff1-
ce00-000000000000
```

4. The previous statements define a partner application (an application that Lync Server 2013 can directly exchange security tokens with) for Exchange Online, which is called microsoft.exchange with the New-CsPartnerApplication cmdlet. If ApplicationIdentifier is different from 00000002-0000-0ff1-ce00-000000000000, the partner application is removed and redefined. If Microsoft.exchange is not defined or is null, the script will create it with the right configuration. Authentication is based on the OAuthServer value that we defined in the first part of the statements. The previously mentioned ApplicationIdentifier is the one we have if we launch only the Get-CsPartnerApplication Microsoft.exchange cmdlet, as shown in the following screenshot:

```
PS C:\Users\administrator.WONDERLAND\Desktop> Get-CsPartnerApplication microsoft
.exchange


Identity                              : microsoft.exchange
AuthToken                             : Microsoft.Rtc.Management.WritableConfig.S
                                        ettings.SSAuth.UseOAuthServer
Name                                  : microsoft.exchange
ApplicationIdentifier                 : 00000002-0000-0ff1-ce00-000000000000
Realm                                 :
ApplicationTrustLevel                 : Full
AcceptSecurityIdentifierInformation   : False
Enabled                               : True
```

5. Run the following statements:

    - Import-Module MSOnlineExtended: To load the Office 365 cmdlets.
    - Connect-MsolService: To connect to Office 365. An administrator's username and password will be required.

6. Now, we need `AppPrincipalId` to retrieve a service principal or a list of service principals from Microsoft Azure AD for Exchange and Lync. The cmdlet to be used is `Get-MsolServicePrincipal`, but it is advisable to save the required output in a text file, in our example, on the `C:\` disk:

   ```
   Get-MsolServicePrincipal | fl >> c:\text.txt
   ```

7. In our scenario, we have (for Exchange and for Lync), as shown in the following screenshot:

   ```
   ExtensionData         : System.Runtime.Serialization.ExtensionDataObject
   AccountEnabled        : True
   Addresses             : {}
   AppPrincipalId        : 00000004-0000-0ff1-ce00-000000000000
   DisplayName           : Microsoft.Lync
   ObjectId              : fb8cc62c-4e76-46f9-83ba-54fad5d14a78
   ServicePrincipalNames : {00000004-0000-0ff1-ce00-000000000000/*.infra.lync.com,
                           00000004-0000-0ff1-ce00-000000000000/*.online.lync.com
                           , 00000004-0000-0ff1-ce00-000000000000, Microsoft.Lync}
   TrustedForDelegation  : True

   ExtensionData         : System.Runtime.Serialization.ExtensionDataObject
   AccountEnabled        : True
   Addresses             : {}
   AppPrincipalId        : 00000002-0000-0ff1-ce00-000000000000
   DisplayName           : Microsoft.Exchange
   ObjectId              : f61a1523-5c97-4369-a5a9-199df3b28344
   ServicePrincipalNames : {00000002-0000-0ff1-ce00-000000000000/outlook.office365
                           .com, 00000002-0000-0ff1-ce00-000000000000/mail.office3
                           65.com,
                           00000002-0000-0ff1-ce00-000000000000/outlook.com,
                           00000002-0000-0ff1-ce00-000000000000/*.outlook.com...}
   TrustedForDelegation  : True
   ```

8. Now, we can use the digital certificate (in X509, the `.cer` format) that we previously generated, and place it in `c:\certificates`, naming it, for example, `Office365.cer`, to run the following cmdlets that will encode the certificate to make it usable to assign to our Office 365 principals:

   ```
   $certificate = New-Object System.Security.Cryptography.X509Certificates.X509Certificate
   $certificate.Import("C:\Certificates\Office365.cer")
   $binaryValue = $certificate.GetRawCertData()
   $credentialsValue = [System.Convert]::ToBase64String($binaryValue)
   ```

9. Now, we have to assign the certificate to the Office 365 principals. The cmdlet that we can use is the following one (for Lync):

   ```
   New-MsolServicePrincipalCredential -AppPrincipalId
   00000004-0000-0ff1-ce00-000000000000 -Type Asymmetric -Usage
   Verify -Value $credentialsValue -StartDate 7/15/2014 -EndDate
   7/3/2015
   ```

> If you see an error like this **New-MsolServicePrincipalCredential : Invalid value for parameter. Parameter name: Credential.EndDate**, it is due to a difference in the time zone between your server and the Office 365 server. Our certificate expiration was on 7/4/2015 (at 01:00 AM), so we have to use 7/3/2015 due to the previously mentioned difference.

10. For Exchange, the statement is:

    ```
    New-MsolServicePrincipalCredential -AppPrincipalId
    00000002-0000-0ff1-ce00-000000000000 -Type Asymmetric -Usage
    Verify -Value $credentialsValue -StartDate 7/15/2014 -EndDate
    7/3/2015
    ```

11. We have to configure Exchange Online Service Principal and configure our on-premises version of Lync Server 2013 as an Office 365 service principal:

    ```
    Set-MSOLServicePrincipal -AppPrincipalID 00000002-0000-0ff1-
    ce00-000000000000 -AccountEnabled $true
    $lyncSP = Get-MSOLServicePrincipal -AppPrincipalID
    00000004-0000-0ff1-ce00-000000000000
    $lyncSP.ServicePrincipalNames.Add("00000004-0000-0ff1-
    ce00-000000000000/lync.contoso.com")
    Set-MSOLServicePrincipal -AppPrincipalID 00000004-0000-0ff1-
    ce00-000000000000 -ServicePrincipalNames $lyncSP.
    ServicePrincipalNames
    ```

There's more...

Doug Deitterick, in a blog post *OAuth Certificate in Lync Server 2013* (http://bit.ly/1l1yWIL), clarified that the OAuth certificate for Lync 2013 is a global one and that it is replicated using the **Central Management Store** (**CMS**). While this kind of replication makes the management of the certificate easier, it also implies that the private key must always be available and that all the Lync Servers have to accept the certification path that released the certificate.

See also

For a deep dive into OAuth and Lync integration, it is useful to watch Bhargav Shukla's presentation at the Microsoft Exchange Conference 2014, *Integrating Exchange 2013 with Lync and SharePoint*, which is available on Channel 9 (http://bit.ly/1rplwvg).

Troubleshooting with client authentication logging

Chapter 12, Lync 2013 Debugging, explores some of the troubleshooting tools commonly used to track a problem on the server side of Lync 2013. However, when talking about authentication (especially when we have an issue that affects a few users), an important source of information is the log that we are able to register on the Lync client. Debugging tools are enabled in different ways on the various devices.

How to do it...

1. For iOS and Android, the option to record information is in the **Advanced Options** that we are able to open from the logon screen. Both operating systems use e-mails to send logs. iOS composes a dedicated message with all the required files attached, while Android attaches a file named `lync_diagnostic.zip`.

2. Windows Phone enables recording from the settings menu. Logs are saved as `.jpg` files in the `Saved Pictures` folder. However, do not send the file using an e-mail because this way, you will lose all the log information. Instead, copy the file on your PC. The desktop client log is enabled in the **Options** | **General** menu. The log files are saved in `c:\users\"username"\AppData\Local\Microsoft\Office\15.0\Lync\Tracing` (`appdata` is a hidden folder).

3. The tool to analyze the previously mentioned logs is called Snooper, an instrument that we will talk about in *Chapter 12, Lync 2013 Debugging*.

See also...

Jason Sloan and Drago Totev, in their blogs, examined the use of Fiddler (a free web debugging proxy) to capture traffic from the mobile clients and debug it. Jason's post *Using Fiddler to troubleshoot Lync Mobile Client* is published at `http://bit.ly/UBCWsg` while Drago's post *Configure Fiddler for Lync Mobile sign-in troubleshooting* is available at `http://bit.ly/1rIaMdU`.

3
Lync Dial Plans and Voice Routing

In this chapter, we will cover the following topics:

- Introducing dial plans and voice routing
- Defining dial plans
- Configuring PSTN usage – voice policy
- Configuring PSTN usage – Location-Based Routing
- Enabling routes
- Validating trunks
- Configuring load balancing, failover, and least cost routing
- Controlling call forwarding

Introduction

Enterprise Voice is a complex group of features in Lync Server 2013, and involves a number of aspects that are as follows:

- Technological, like the architecture of our deployment and our hardware
- Economical, like Lync user licenses or Office 365 fees
- Logistical, like the agreements with one or more external providers
- Legal, like compliance with local regulations

Lync Dial Plans and Voice Routing

This chapter focuses on some specific aspects of the flow used to manage a voice call from a user in Lync Server 2013. We can imagine the dialing of a phone number divided in two different phases. The first one is usually called dialing behaviors and has only one objective, that is, to normalize any dialed number until it is compatible with E.164 (or to apply a less used but more precise definition, the RFC3966 standard). Dial plans and normalization rules are used during this phase. The second phase is dedicated to a series of controls to resolve if a call is authorized and what is the best path for it. This is usually called the routing and authorization phase. Voice policies, voice routes, **Public Switched Telephone Network** (**PSTN**) usage, and trunk configuration are applied during this stage. We will see the configurations and hints related to both the previously mentioned stages of call processing in Lync.

Introducing dial plans and voice routing

The base condition for a Lync Enterprise Voice deployment to work correctly is to have all telephone numbers formatted in the E.164 standard. The previously mentioned standard requires a plus sign, followed by a maximum of three digits dedicated to **Country Code** (**CC**), and the remaining numbers dedicated to national standards, in order to identify subscribers. Depending on the length of the CC, the part that contains the city (area) code, local number, and extensions might vary (the maximum total length, anyway, is 15 digits, excluding extensions). Each country is able to decide how many digits it should use (considering that a 15-digit number allows for 100 trillion different permutations). The *ITU E.164* documentation about the number structure for geographic areas (at http://www.itu.int/rec/T-REC-E.164-201011-I/en) divides the number in CC (1-3 digits), **National Destination Code** (**NDC**), and **Subscriber Number** (**SN**). NDC and SN are collectively called the national (significant) number. The schema is shown in the following image:

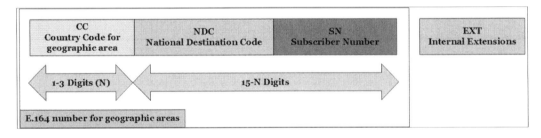

Normalization (converting numbers dialed in various formats to the E.164 format) is important for many reasons, including the fact that users in different locations are able to dial our phone numbers correctly only if they stick to the E.164 format. Dial plans in Lync contain rules that are used to normalize phone numbers, starting from the way our users are comfortable in dialing them. Normalization is performed using normalization rules associated with Lync dial plans. In the following screenshot, we can see the rules inside a dial plan:

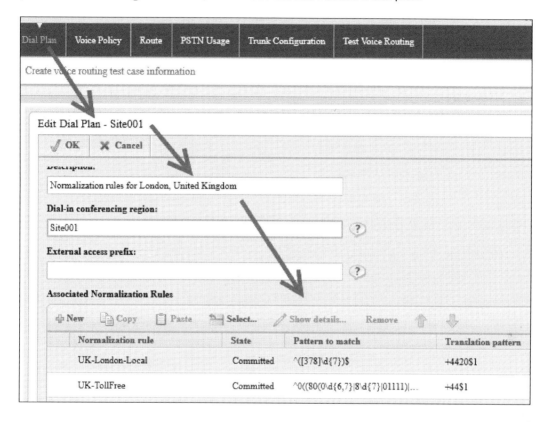

Dial plans are applied starting from the global level and descending to the site level, then to the pool level, and lastly to the user level. The last policy plan in the aforementioned list is the only one that is applied, so that we are able to manage all kinds of exceptions. Enterprise Voice requires additional configuration, including the following:

- Voice policies, which are used to control the kind of calls the user is enabled to perform (local, international, and so on). Voice policies in Lync determine whether a user is permitted to successfully make a call using the dialed number or not.
- Voice routes, which are required to determine whether there is a regular expression that matches the call and routes it to the appropriate gateway.

Lync Dial Plans and Voice Routing

- PSTN usage, which is essential to associate routes and voice policies and to verify which users are allowed to use specific routes (like the one we can see in the following screenshot):

- Trunk configuration, which is used to manage translation rules and communication with our gateways (or with the ones of our ITSP).

> In the past, it was a standard for companies that use a **Private Branch Exchange** (**PBX**) to rely on the behaviors and dialing habits of the users. A typical example was the requirement for digit 0 to keep the external line. The previously mentioned scenario changes completely with Lync Server 2013, which is based on a series of rules to normalize every telephone number to the E.164 standard with no user intervention required.

Getting ready

To define the base policies, we will use the Lync Dial Rule Optimizer tool by Ken Lasko. This tool provides great help in defining some voice-configuration aspects. In this first recipe, we will provide a quick overview of the tool and use specific parts to understand the logic behind a Lync voice configuration. We will come back to the tool in more detail in *Chapter 5, Scripts and Tools for Lync*.

Chapter 3

How to do it...

1. We will use a medium-sized office in London in the example. The office will have a range of phone numbers that start from +442077030001 and end at +442077030099. The E.164 logic is the one shown in the following image:

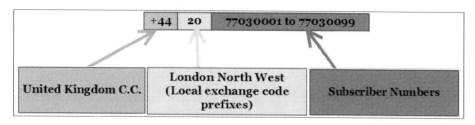

2. It is important to save our existing rules. We can do it by using the `Export-CsConfiguration -FileName "C:\Backup\LyncConfigBackUp.zip"` cmdlet for a complete backup of our Lync Server configuration. However, if we want to take a backup of only specific parts of the deployment, we can use the cmdlets shown in *Chapter 10, Managing Lync 2013 Backup and Restore*, in the *Configuration information* recipe.

3. A PSTN gateway must exist for the site where we are going to apply the rules generated with the Optimizer (as shown in the following screenshot from the Lync Topology Builder):

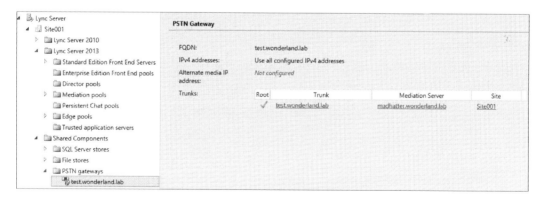

Lync Dial Plans and Voice Routing

4. The next step is to insert our information in the Optimizer page that will generate the `.ps1` file required to create all the necessary rules and policies, as shown in the following screenshot:

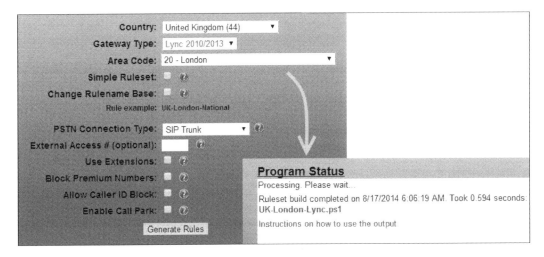

5. The script (in our example, `UK-London-Lync.Ps1`) when running, will just require the Site ID and a selection between the creation of user-level dial plans or site-level dial plans. Dial plan, voice policies, routes, PSTN usages, and trunk configuration are added automatically for the site, as shown in the following screenshot:

```
PS C:\> .\UK-London-Lync.ps1

SiteId Identity
------ --------
1      Site:Site001

Enter the Site ID to apply the dialing rules for London, United Kingdom: 1

Create site-level or user-level dial plan?
[S] Site  [U] User  [?] Help (default is "S"): s

Creating site-level dial plan
Creating normalization rules
Creating voice policies
Creating PSTN usages
Assigning usages to voice policies
Creating voice routes
Creating outbound translation rules
Finished!
PS C:\>
```

Chapter 3

How it works...

The script itself is a complex one (and contains currently more than 700 lines). Part of the complexity is because a great number of parameters are controlled, adding flexibility to the result. It is also possible to manage the different options outside the user interface, running the following cmdlet with optional parameters (there is one example included inside the script, which shows how to apply least cost routing):

```
.\UK-London-Lync.ps1 -SiteID 2 -DialPlanType User
-LeastCostRouting:$TRUE -LCRSites 'US-NY-NewYorkCity,UK-London,SG-
Singapore'
```

See also

Talking about dial plans and voice routing, I suggest that you read an interesting post by Chris Williams, *The Difference Between Dial Plans and Voice Routes* at http://www.lyncinsider.com/lync-server-2013/the-difference-between-dial-plans-and-voice-routes/, which clarifies the difference between the two aspects of voice processing in Lync.

Another post that I suggest you read is *Demystify Lync Enterprise Voice Phone Numbers and Extension* by Thomas Poett at http://lyncuc.blogspot.it/2013/02/demystify-lync-enterprise-voice-phone.html. Thomas' work examines many aspects of Enterprise Voice, including phone number management and E.164 normalization.

Defining dial plans

As we mentioned before, for Enterprise Voice, Lync requires a dial plan that contains normalization rules that elaborate the phone number typed in by the user and standardizes it to the E.164 format. Normalization rules use .NET regular expressions (http://msdn.microsoft.com/en-us/library/hs600312.aspx).

Getting ready

As a starting point, I have used the `UK-London-Lync.ps1` script generated with Ken Lasko's Lync Dialing Rule Optimizer.

How to do it...

1. To generate normalization rules, it is necessary to have at least an existing dial plan. We can create a site-level dial plan with the following cmdlet:

   ```
   New-CsDialPlan -identity Site:Site001
   ```

2. The normalization rules created with the following cmdlets will be associated with the previously mentioned dial plan:

   ```
   New-CsVoiceNormalizationRule -Name "UK-London-Local" -Parent "Site:Site001" -Pattern '^([378]\d{7})$' -Translation '+4420$1' -Description "Local number normalization for London, United Kingdom"
   New-CsVoiceNormalizationRule -Name 'UK-TollFree' -Parent "Site:Site001" -Pattern '^0((80(0\d{6,7}|8\d{7}|01111)|500\d{6}))$' -Translation '+44$1' -Description "TollFree number normalization for United Kingdom"
   New-CsVoiceNormalizationRule -Name 'U K-Premium' -Parent "Site:Site001" -Pattern '^0((9[018]\d|87[123]|70\d)\d{7})$' -Translation '+44$1' -Description "Premium number normalization for United Kingdom"
   New-CsVoiceNormalizationRule -Name 'UK-Mobile' -Parent "Site:Site001" -Pattern '^0(7([1-57-9]\d{8}|624\d{6}))$' -Translation '+44$1' -Description "Mobile number normalization for United Kingdom"
   New-CsVoiceNormalizationRule -Name 'UK-National' -Parent "Site:Site001" -Pattern '^0((1[1-9]\d{7,8}|2[03489]\d{8}|3[0347]\d{8}|5[56]\d{8}|8((4[2-5]|70)\d{7}|45464\d)))(\D+\d+)?$' -Translation '+44$1' -Description "National number normalization for United Kingdom"
   New-CsVoiceNormalizationRule -Name 'UK-Service' -Parent "Site:Site001" -Pattern '^(1(47\d|70\d|800\d|1[68]\d{3}|\d\d)|999|[\*\#][\*\#\d]*\#)$' -Translation '+$1' -Description "Service number normalization for United Kingdom"
   New-CsVoiceNormalizationRule -Name 'UK-International' -Parent "Site:Site001" -Pattern '^(?:00)((1[2-9]\d\d[2-9]\d{6})|([2-9]\d{6,14}))(\D+\d+)?$' -Translation '+$1' -Description "International number normalization for United Kingdom"
   ```

3. Now, we can look at some of the rules to understand the logic:

   - We can think of rules being made up by a **Pattern** part and a **Translation** part. The pattern contains designators and variables that represent a specific set and quantity of numbers; for example, in the **UK-London-Local** rule, (`'^([378]\d{7})$'`) is a pattern that matches any phone number that starts with 3,7, or 8 (and the following 7 digits). Then, for the translation, the $ symbol captures the digits from the pattern (which were included inside the parentheses) and normalizes them into the E.164 format prefix with +4420 (`'+4420$1'`). The rule normalizes local calls in London; for example, a call to 70405060 will be translated to +44 20 70405060.

- The **UK-TollFree** rule (`'^0((80(0\d{6,7}|8\d{7}|01111)|500\d{6}))$'`) standardizes to E.164 tool-free numbers in the UK (starting with 0800, 0808, and 0500), removing the initial 0 and adding the CC of United Kingdom, +44 (`'+44$1'`).

In the following image, we can see a schema of the phone numbers that we are going to manage with this rule:

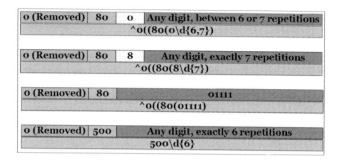

- The **UK-Premium** rule (`'^0((9[018]\d|87[123]|70\d)\d{7})$'`) standardizes numbers that are mainly used for horoscopes, chat lines, TV voting and other services (090, 091, and 098), customer service lines (0871, 0872, and 0873), and numbers used to divert calls to another phone number (070). Again, the initial 0 is removed and the CC of United Kingdom, +44, is added (`'+44$1'`).

- The **UK-International** rule (`'^(?:00)((1[2-9]\d\d[2-9]\d{6})|([2-9]\d{6,14}))(\D+\d+)?$'`) normalizes international calls (the first half of the rule takes care of North American numbers, and the second half takes care of the rest of the world). The rule acts by removing the initial 00 and capturing all the calls to numbers with a pattern like the one in the following schema (the + sign is added, `'+$1'`):

00 (Removed)	1	One digit, between 2 and 9	One digit	One digit	One digit, between 2 and 9	Any digit, exactly 6 repetitions
00 (Removed)	One digit, between 2 and 9				Any digit, between 6 and 14 repetitions	

Lync Dial Plans and Voice Routing

There's more...

To read (and edit) the regular expressions used in Lync normalization rules, I suggest that you use a free tool, Expresso 3.0 (`http://www.ultrapico.com/ExpressoDownload.htm`). As we can see in the following screenshot, using this tool makes it easier to understand complex regex strings, like the ones we need to manipulate phone numbers:

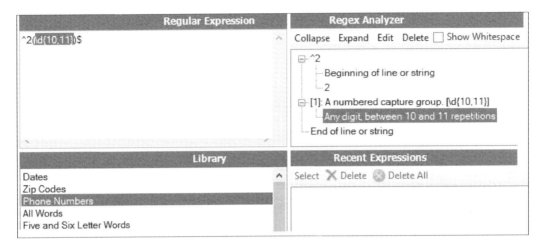

Please note that Expresso requires a (free) registration to keep using it after 60 days.

Configuring PSTN usage – voice policy

With PSTN usage records, we specify what is managed as a local call, a long distance call, an international call and so on. We must connect PSTN usage records with voice policies (assigned to users) and routes (assigned to phone numbers). What we are going to see now is a possible sequence of steps to configure voice policies and PSTN usages. However, there is no fixed order in Lync, so there is no single way to organize the various aspects of Enterprise Voice. Microsoft TechNet offers some information, starting with the *Deploying Enterprise Voice in Lync Server 2013* post at `http://technet.microsoft.com/en-us/library/gg412876.aspx`. The suggested order in the previously mentioned document is as follows:

- Deploying mediation servers
- Configuring trunks with PSTN gateways, IP-PBXs, or using SIP with an **Internet Telephony Service Provider** (**ITSP**)
- Defining dial plans

- Configuring the following parameters:
    - Voice policies
    - PSTN usage records
    - Voice routes
- Integrating with Exchange
- Deploying advanced voice features such as call admission control, emergency services (E9-1-1), media bypass, and Location-Based Routing

Another aspect to consider is that there is no predefined naming convention. A recommended approach could be the following one, suggested by Ken Lasko:

- PSTN usage names relate to calls of the same type, such as local, long distance/national, or international. If we have multiple sites, it is useful to add site information in the country-state/prov-city-callclass or country-city-callclass (for example, **UK-London-Local**) format. We can use the same naming convention for voice policies.
- Routes: at least five for each gateway, that is, local, national (long distance), international, toll-free, and service codes (for example, **UK-London-TollFree**).
- National and international routes will be added into the national and international PSTN usages respectively. Local, toll-free into the local PSTN usage. It is a good idea to add service codes into a dedicated PSTN usage (service PSTN usage). Service numbers generally don't work outside of the country, and this could result in unintended behavior when using least-cost routing.
- Normalization rules: country-state/prov-city-routeclass or country-city-routeclass.

How to do it...

1. We will start with creating empty voice policies that are not associated with PSTN usage records. The logic here is to have three different profiles of authorization for our Lync users: local calls, national calls, and international calls. The default policy for our site (`site:site001`) is the last one we are going to create, and it allows local, national, and mobile calls:

    ```
    New-CSVoicePolicy "UK-London-Local" -Description "Allows local
    calls from London, United Kingdom"
    New-CSVoicePolicy "UK-London-National" -Description "Allows local/
    national calls from London, United Kingdom"
    New-CSVoicePolicy "UK-London-International" -Description "Allows
    local/national/international calls from London, United Kingdom"
    New-CSVoicePolicy site:site001 -Description "Allows local/national
    calls from London, United Kingdom"
    ```

Lync Dial Plans and Voice Routing

2. For every policy, we will receive the following screentext: **WARNING: No PSTN Usage specified. Users granted this Voice Policy will not be able to make outbound PSTN calls**. We will populate them later.

3. Now, we have to define PSTN usage policies. The logic will be the same that we used for voice policies, with different policies to identify different levels of service:

   ```
   Set-CsPSTNUsage -Identity global -Usage @{Add="UK-London-Local"}
   Set-CsPSTNUsage -Identity global -Usage @{Add="UK-London-Service"}
   Set-CsPSTNUsage -Identity global -Usage @{Add="UK-London-National"}
   Set-CsPSTNUsage -Identity global -Usage @{Add="UK-London-Mobile"}
   Set-CsPSTNUsage -Identity global -Usage @{Add="UK-London-Premium"}
   Set-CsPSTNUsage -Identity global -Usage @{Add="UK-London-International"}
   ```

   Again, our PSTN usage policies are empty and will be populated later.

4. Now, it is possible to associate the voice policies and the PSTN usage policies. We will start with the default voice policy associated with the site:

   ```
   Set-CsVoicePolicy -Identity site:site001 -PstnUsages UK-London-Local
   Set-CsVoicePolicy -Identity site:site001 -PstnUsages UK-London-Mobile
   Set-CsVoicePolicy -Identity site:site001 -PstnUsages UK-London-National
   Set-CsVoicePolicy -Identity site:site001 -PstnUsages UK-London-Service
   ```

5. Then, we will associate PSTN usage policies with the remaining voice policies, starting with the one that authorizes international/premium calls:

   ```
   Set-CsVoicePolicy -Identity UK-London-International -PstnUsages UK-London-Local
   Set-CsVoicePolicy -Identity UK-London-International -PstnUsages UK-London-Mobile
   Set-CsVoicePolicy -Identity UK-London-International -PstnUsages UK-London-National
   Set-CsVoicePolicy -Identity UK-London-International -PstnUsages UK-London-Service
   Set-CsVoicePolicy -Identity UK-London-International -PstnUsages UK-London-Premium
   Set-CsVoicePolicy -Identity UK-London-International -PstnUsages UK-London-International
   ```

6. Let's proceed with the authorization policy for national and local calls:

   ```
   Set-CsVoicePolicy -Identity UK-London-National -PstnUsages UK-London-Local
   Set-CsVoicePolicy -Identity UK-London-National -PstnUsages UK-London-Mobile
   Set-CsVoicePolicy -Identity UK-London-National -PstnUsages UK-London-National
   Set-CsVoicePolicy -Identity UK-London-National -PstnUsages UK-London-Service
   Set-CsVoicePolicy -Identity UK-London-Local -PstnUsages UK-London-Local
   Set-CsVoicePolicy -Identity UK-London-Local -PstnUsages UK-London-National
   ```

7. The following cmdlets will perform the last step, generating voice routes and associating them with PSTN usage records. The scenario that we are using has a single PSTN gateway. If there are multiple gateways, the rules have to be modified accordingly:

   ```
   New-CSVoiceRoute -Name "UK-London-Local" -Priority 0 -PSTNUsages "UK-London-Local" -PSTNGatewayList test.wonderland.lab -NumberPattern '^\+4420([378]\d{7})$' -Description "Local routing for London, United Kingdom"
   New-CSVoiceRoute -Name "UK-London-Mobile" -Priority 2 -PSTNUsages "UK-London-Mobile" -PSTNGatewayList test.wonderland.lab -NumberPattern '^\+447([1-57-9]\d{8}|624\d{6})$' -Description "Mobile routing for London, United Kingdom"
   New-CSVoiceRoute -Name "UK-London-TollFree" -Priority 3 -PSTNUsages "UK-London-Local" -PSTNGatewayList test.wonderland.lab -NumberPattern '^\+44(80(0\d{6,7}|8\d{7}|01111)|500\d{6})$' -Description "TollFree routing for London, United Kingdom"
   New-CSVoiceRoute -Name "UK-London-Premium" -Priority 4 -PSTNUsages "UK-London-Premium" -PSTNGatewayList test.wonderland.lab -NumberPattern '^\+44(9[018]\d|87[123]|70\d)\d{7}$' -Description "Premium routing for London, United Kingdom"
   New-CSVoiceRoute -Name "UK-London-National" -Priority 5 -PSTNUsages "UK-London-National" -PSTNGatewayList test.wonderland.lab -NumberPattern '^\+44(1[1-9]\d{7,8}|2[03489]\d{8}|3[0347]\d{8}|5[56]\d{8}|8((4[2-5]|70)\d{7}|45464\d))$' -Description "National routing for London, United Kingdom"
   New-CSVoiceRoute -Name "UK-London-International" -Priority 7 -PSTNUsages "UK-London-International" -PSTNGatewayList test.wonderland.lab -NumberPattern '^\+((1[2-9]\d\d[2-9]\d{6})|(?(?!(44))([2-9]\d{6,14})))$' -Description "International routing for London, United Kingdom"
   New-CSVoiceRoute -Name "UK-London-Service" -Priority 6 -PSTNUsages "UK-London-Service" -PSTNGatewayList test.wonderland.lab -NumberPattern '^\+(1(47\d|70\d|800\d|1[68]\d{3}|\d\d)|999|[\*\#][\*\#\d]*\#)$' -Description "Service routing for London, United Kingdom"
   ```

There's more...

In the various steps, to keep the logic as clear as possible, I have used a cmdlet for every single configuration. A more compact form is the one that uses the @ operator. For example, we can define the `International` voice policy with the following single cmdlet:

```
Set-CsVoicePolicy -Identity UK-London-International -PstnUsages @
{add="UK-London-Local", "UK-London-Mobile", "UK-London-National", "UK-
London-Service", "UK-London-Premium", "UK-London-International"}
```

See also

Yoav Barzilay has published an interesting script that allows assigning Lync policies (with a user scope, including voice policies) to Lync-enabled users based on their membership in an Active Directory group. The script requires administrative rights to run. The Lync Management Shell and the Active Directory PowerShell module are required on the computer where the script will run. To download the script and know more about it, please refer to this post *Assign Lync Policies To Lync Users Based On Active Directory Group Membership* at `http://y0av.me/2014/07/07/adpt/`.

Configuring PSTN usage – Location-Based Routing

Getting ready

Location-Based Routing (**LBR**) is a feature of Enterprise Voice that routes calls based on user location instead of **Least Cost Routing** (**LCR**). The advantage of LCR is related to costs because Lync will use the gateway that is closest to the called party (using the corporate wide area network, WAN, to reach to the previously mentioned gateway). This kind of solution minimizes tool charges. However, some reasons (the desire to minimize WAN utilization or local regulations) could require LBR to route the call to the gateway that is physically nearer to the Lync user. Lync 2013 CU1 is at least required to enable LBR. Lync Enterprise Voice configuration is a prerequirement for LBR. It is also required to configure at least a network region (a backbone or a hub in our network), a site (a geographical location, such as a branch office), and a subnet (based on the IP subnet). LBR is based on voice routing policies associated with network regions, so the previously mentioned configuration is mandatory.

How to do it...

1. We will define Lync regions, sites, and subnets (in this order). We can use Lync Management Shell or the Lync Control Panel, as shown in the following screenshot:

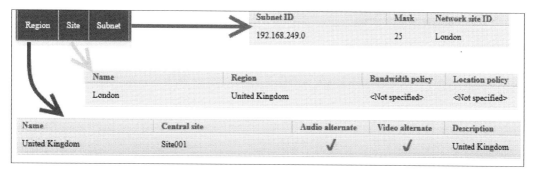

2. A voice routing policy must associate the network site with the appropriate PSTN usages (only **UK-London-International** is excluded because we are not going to use LBR with this kind of call):

   ```
   New-CsVoiceRoutingPolicy -Identity "LBR_London" -Name "Location
   Based Routing London" -PstnUsages @{add="UK-London-Local", "UK-
   London-Mobile", "UK-London-National", "UK-London-Service", "UK-
   London-Premium"}
   ```

   We can use the `Get-CsNetworkSite` cmdlet to get the site ID for the existing sites. The base syntax of the cmdlet is `Set-CsNetworkSite -Identity <site ID> -EnableLocationBasedRouting $true -VoiceRoutingPolicy <voice routing policy ID>`, in our example, `Set-CsNetworkSite -Identity London -EnableLocationBasedRouting $true -VoiceRoutingPolicy Tag:LBR_London`.

3. On the single trunk where we want to enforce routing restrictions, we have to use a cmdlet like the following one:

   ```
   Set-CsTrunkConfiguration -Identity PstnGateway:test.wonderland.lab
   -EnableLocationRestriction $true -NetworkSiteID London
   ```

4. Lync Server 2013 includes a feature called the toll bypass capability. To bypass long-distance toll charges, we can redirect voice calls to use a PSTN gateway that is closest to the location of the person being called. If we want to keep routing restrictions in place, the voice policy must prevent PSTN toll bypass. The cmdlet to use for the various policies is similar to the following one:

   ```
   Set-CsVoicePolicy -Identity "UK-London-National"
   -PreventPSTNTollBypass $true
   ```

> Although the steps we have seen in this recipe are the ones suggested in TechNet, Ken Lasko suggests that the `EnableLocationRestriction:$TRUE` parameter should only be used to stop inbound PSTN calls while a user is at a site where call routing restrictions apply. His suggestion is to bypass the `Set-CSTrunkConfiguration -EnableLocationRestriction:$TRUE` parameter to enable location-based routing without inbound call restrictions (http://ucken.blogspot.ca/2013/06/location-based-routing-without-inbound.html).

There's more...

A session of the Lync Conference 2014 is dedicated to *Location Based Routing* (http://channel9.msdn.com/Events/Lync-Conference/Lync-Conference-2014/VOICE303). It is an interesting and intensive deep dive on this topic.

An overview of *Lync Enterprise Voice Core Infrastructure Updates* is part of Lync 2013 Ignite that is available at http://channel9.msdn.com/Series/Lync-2013-Ignite/Lync-Enterprise-Voice-Core-Infrastructure-Updates.

Enabling routes

In Lync Server 2013, Microsoft has added the capability to associate the following:

- A single Mediation Server for multiple gateways
- A single gateway for multiple Mediation Servers
- A single gateway multiple times with the same Mediation Server

This feature is called **many-to-many** (**M:N**) trunk routing, and it allows Mediation Servers to handle multiple logical paths (trunks) from a gateway on multiple different ports. M:N routing adds redundancy and flexibility to Lync deployments. In the following steps, we will see how to enable multiple gateways for a voice route (adding redundancy) and how to use multiple connections to a single gateway to add flexibility in the trunk configuration.

How to do it...

1. Our initial scenario will include two PSTN gateways (**test.wonderland.lab** and **2ndtest.wonderland.lab**) and a single Mediation Server (**madhatter.wonderland.lab**). Lync Topology Builder will automatically create two trunks that connect the single Mediation Server to the gateways, as shown in the following screenshot:

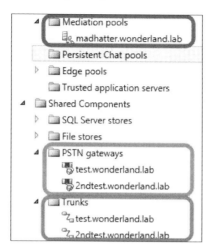

2. Using the Lync Control Panel, we are able to use both the trunks inside a voice route to add redundancy, as shown in the following screenshot for the **UK-London-National** route:

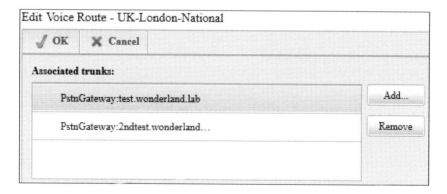

Note that load balancing, failover, and least cost routing are explained in a dedicated recipe later in this chapter.

Lync Dial Plans and Voice Routing

3. We have a requirement from our company to always have SRTP encryption on our communication with the gateway (only for the UK-London-Premium voice route). This is something we can achieve by configuring an additional trunk between the Mediation Server and a gateway. To enable multiple trunks from a Mediation Server to a gateway, we have to add listening ports on the server configuration in the Lync Topology Builder, as shown in the following screenshot:

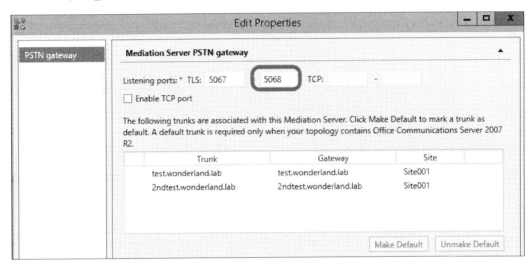

4. It is now possible to define a new trunk from our Mediation Server to the **2ntest.wonderland.lab** gateway (we will call it `2ndtest_encryption_trunk`), using the Topology Builder, as we can see in the following screenshot:

5. In the Lync Control Panel, we will configure the additional trunk to always require encryption, as shown in the following screenshot:

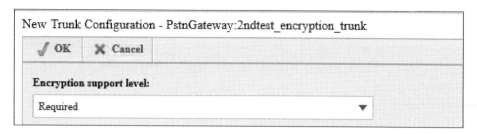

6. Associating the voice route with the **2ndtest_encryption_trunk** will satisfy the SRTP encryption requirement. In the following screenshot, the details of the configuration performed in the Lync Control Panel is shown:

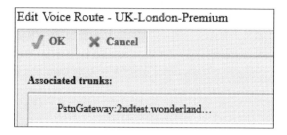

There's more...

By default, trunks in Lync 2013 have a parameter called `EnableFastFailoverTimer` set to `True`. With this configuration, the gateway must respond in 10 seconds or the call will be routed to a different gateway. Although this is usually not an issue, some gateways might be slow in completing the call, generating multiple events such as **46046, LS Outbound Routing** and **Cause: All gateways available for this call are marked as down**. A real-world example could be a sip trunk that makes a call to the mobile network; sometimes it takes more than 10 seconds and the call is dropped. The failover timer can be disabled with the following cmdlet `Set-CsTrunkConfiguration -Identity Service:PstnGateway:test.wonderland -EnableFastFailoverTimer $False` (we are using the `test.wonderland.lab` gateway), or from Control Panel, as shown in the following screenshot:

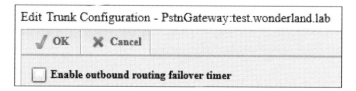

> It is always recommended to first take a look at this kind of issue and try to fix it. Disabling the `EnableFastFailoverTimer` parameter can have a negative impact on the system.

Validating trunks

In Lync 2013, trunk configuration settings define a relationship between Mediation Server and PSTN gateway, **IP-public branch exchange** (**PBX**), or **Session Border Controller** (**SBC**) in the service provider. There are cmdlets to create, modify, and delete trunks. In addition, there is a cmdlet `Test-CsTrunkConfiguration`, which allows us to validate a trunk configuration against a phone number. The previously mentioned cmdlet is part of a particular category of commands, and is known as Lync Synthetic Transactions, which enables testing core functionality by simulating interaction between users, computers, and so on, without having to fire up test workstations with the client installed.

How to do it...

1. We can test `2ndtest_encryption_trunk`, which we defined in the previous section. Its characteristic is that it has the `SRTPMode` parameter set to `Required`.
2. The synthetic test will require the trunk configuration. The easiest way to manage this is to save the trunk configuration information in a variable as follows:

   `$tc = Get-CsTrunkConfiguration -Identity Service:PstnGateway:2ndtest_encryption_trunk`

3. The last step is to launch the test that verifies a phone number to dial `Test-CsTrunkConfiguration -dialednumber +44981234567 -TrunkConfiguration $tc`.

There's more...

We are able to check whether the current user is enabled to use the `Test-CsTrunkConfiguration` synthetic test with the `Get-CsAdminRole | Where-Object {$_.Cmdlets -match "Test-CsTrunkConfiguration"}` cmdlet that verifies the **Role-Based Access Control** (**RBAC**) permissions.

Chapter 3

See also

Murali Krishnan has published a script in his blog that automatically performs many synthetic tests for a single pool deployment at `http://unifiedme.co.uk/2013/11/lync-server-2013-synthetic-test-script/`. The Topology Validator in the Lync 2010 resource kit is able to (also) perform more than 40 synthetic tests, but there is a similar tool in the Lync 2013 version of the resource kit (`http://www.microsoft.com/en-us/download/details.aspx?id=36821`).

Configuring load balancing, failover, and least cost routing

In a previous recipe, we discussed M:N trunk routing, and we said that associating multiple gateways with a single Mediation Server adds resiliency and flexibility. In this recipe, we will see how to obtain three different results: load balancing, failover, and least cost routing.

> There are some limitations and compatibility requirements while implementing the solutions explained in this recipe. We can find a short explanation about them in the TechNet post *M:N trunk in Lync Server 2013* at `http://technet.microsoft.com/en-us/library/gg398971.aspx`.

Getting ready

In the following recipe, we will add some gateways to our scenario and we will assume that they are already configured. The **international.wonderland.lab** gateway is the least costly gateway for international calls from UK to France. To enable failover routing, we will add another gateway **failover.wonderland.lab**.

How to do it...

1. To configure load balancing, the first step is to take a list of the routes associated with a specific voice policy. For example, UK-London-National, which we created in a previous recipe, is associated with the following PSTN usage records: UK-London-Local, UK-London-Mobile, UK-London-National, and UK-London-Service. They have associated routes with the same name.

Lync Dial Plans and Voice Routing

2. We are required to define the PSTN gateways that we want to use to balance in each voice route. In the following screenshot, we are load balancing the **UK-London-Mobile** voice route:

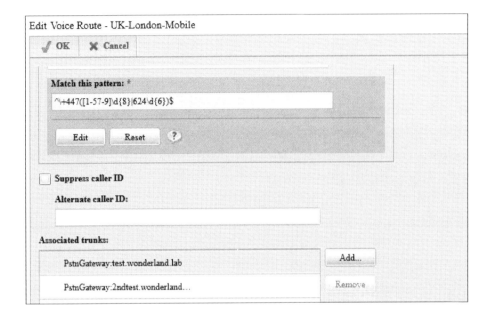

Lync will load balance voice traffic using a round robin between the PSTN gateways defined in the various voice routes.

3. As we mentioned in the *Getting ready* section of this recipe, we will add a PSTN gateway to our scenario, **international.wonderland.lab**, which is the least costly gateway for international calls from UK to France (the country code is +33).

4. To enable least cost routing, we have to define a PSTN usage rule and associate it with a voice route that has to match calls to the CC +33. The route must be connected to the **international.wonderland.lab** gateway.

5. With a cmdlet that we already used in the *Configuring PSTN usage – voice policy* recipe, we can define a PSTN usage record called `UK-International-France`:

   `Set-CsPstnUsage -Identity global -Usage @{add="UK-International-France"}`

6. Using the Control Panel (as an alternative to the commands that we used previously in the chapter), we have to define a new voice route with the following parameters:

   - Matching with the `^(?:00)(33\d{9-10})` normalization rule
   - Associated with the trunk that connects to the `international.wonderland.lab` gateway
   - Associated with the UK-International-France PSTN usage rule

Chapter 3

The previously mentioned parameters are shown in the following screenshot:

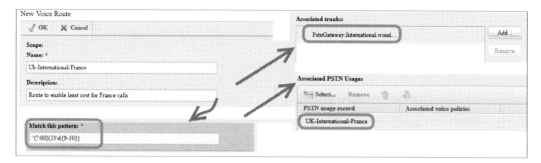

7. After editing the **UK-London-International** voice policy so that the **UK-International-France** PSTN usage record is located above **UK-London-International**, we are sure that all the calls to a +33 phone number will use the least costly route. The configuration is the one shown in the following screenshot:

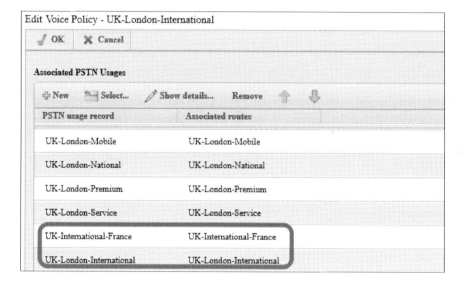

8. To enable a failover route, we are required to edit voice policies so that we have more than one PSTN usage rule inside the single policy, and associated with different voice routes. If we have two voice routes available, the second one will be used as a failover. We will proceed to define a failover for the **UK-London-National** voice policy.

9. We will define a new PSTN usage record **UK-Failover**:
   ```
   Set-CsPstnUsage -Identity global -Usage @{add="UK-Failover"}
   ```

Lync Dial Plans and Voice Routing

Now, we will perform the following steps:

- A new voice route is required. We will call it `UK-National-Failover`. It will be associated with the same normalization rules of the **UK-London-National** route.
- Use the trunk to the **failover.wonderland.lab** gateway.
- Be associated with the **UK-Failover** PSTN usage rule.

The following image highlights the two voice routes:

10. After editing the **UK-London-National** voice policy (for example) and applying the PSTN usage rules in the correct order, we defined a failover for the preferred voice route, as shown in the following screenshot:

There's more...

Ken Lasko has dedicated a series of posts to Lync Enterprise Voice Misconceptions that also include important information about failover routing at `http://ucken.blogspot.ch/2014/09/EVMisconceptionsRouting.html`.

Keenan Crockett, in a blog post dedicated to Lync Server 2010, *Distinguishing Between Voice Routes Configured for Load Balancing vs. Voice Policies Configured for Failover and Least Cost Routing in Lync Server 2010* (`http://blogs.perficient.com/microsoft/2011/08/distinguishing-between-voice-routes-configured-for-load-balancing-vs-voice-policies-configured-for-failover-and-least-cost-routing-in-lync-server-2010/`) outlined some scenarios for the different configurations, also pointing out some wrong ways to realize them.

Chapter 3

Controlling call forwarding

Call forwarding is a feature usually managed on the client side of Lync. If call-forwarding options are enabled, users can forward phone calls to another number or to another contact. The Set-CsVoicePolicy and Get-CsVoicePolicy cmdlets enable the Lync administrator to centrally manage this feature.

How to do it...

1. Our company's requirement is to disable call forwarding for the users that are included in the **UK-London-Local** voice policy. The first step is to verify which voice policies enable the feature. We can use the following statement:

   ```
   Get-CsVoicePolicy | Where-Object {$_.AllowCallForwarding -eq $True}
   ```

2. The **UK-London-Local** policy allows all forwarding. To disable it, we can use the following command:

   ```
   Set-CsVoicePolicy UK-London-Local -Allowcallforwarding $false
   ```

There's more...

It is possible to manage Lync Call Forwarding using the **Secondary Extension Feature Activation Utility** (**SEFAUtil**), which is part of the Lync Server 2013 resource kit tools (an explanation about deploying SEFAUtil is available on the Microsoft TechNet page at http://technet.microsoft.com/en-us/library/jj945659.aspx). Andrew Price outlined the process in a post *Call Forwarding via SEFAUtil* (http://lyncme.co.uk/microsoft-lync-server-2010/call-forwarding-via-sefautil/) on his blog.

SEFAUtil allows us to manage features such as simultaneous ringing, unanswered timeout forwarding, and call pickups (with Lync Server 2013).

4
Lync 2013 Integration with Exchange

In this chapter, we will cover the following recipes:

- Configuring the Unified Messaging integration
- Configuring OAuth and partner applications for Lync 2013 and Exchange 2013 integration
- Configuring OAuth between Lync 2013 and Exchange 2013
- Configuring Lync 2013 and Exchange 2013 as partner applications
- Configuring Lync 2013 to use Exchange 2013 for archiving
- Configuring Lync 2013 to use the Exchange 2013 Unified Contact Store
- Integrating Lync 2013 with Exchange 2013 Outlook Web App

Introduction

In this chapter, we will describe the levels of integration between Lync 2013 and Exchange 2013 and 2010. We will highlight the differences depending on the Exchange version and describe the procedures to configure and test this integration.

Levels of integration between Lync 2013 and Exchange

Lync 2013 and Exchange can integrate at several levels to provide the end user with the best experience possible. The integration is done both at the client and server levels.

To fully integrate Lync and Exchange, there are several tasks that need to be done at the server level, using the following tools:

- The Lync 2013 Management Shell
- The Exchange 2013 or 2010 Management Shell
- The Lync 2013 Control Panel
- The Exchange 2013 Administrative Centre or the Exchange 2010 Management Console
- The Lync 2013 OCSUmUtil tool

These tools will allow you to configure the integration between Lync 2013 and Exchange. For official guidance on how to integrate both platforms, refer to the Microsoft Lync 2013 Technet article at http://technet.microsoft.com/en-us/library/jj688098.aspx.

Depending on the Lync and Exchange versions, the integration features available will be different, and the following table highlights these differences:

Lync Version	Exchange 2013	Exchange 2010
Lync 2013	<ul><li>Exchange Unified Messaging</li><li>Outlook Web App Instant Messaging and Presence</li><li>Unified Contact Store</li><li>High-resolution photos</li><li>Outlook Web App Meeting Scheduling</li><li>Instant Messaging Archiving</li></ul>	<ul><li>Exchange Unified Messaging</li><li>Outlook Web App Instant Messaging and Presence</li></ul>
Lync 2010	<ul><li>Exchange Unified Messaging</li><li>Outlook Web App Instant Messaging and Presence</li></ul>	<ul><li>Exchange Unified Messaging</li><li>Outlook Web App Instant Messaging and Presence</li></ul>
OCS 2007 R2	<ul><li>Exchange Unified Messaging</li><li>Outlook Web App Instant Messaging and Presence</li></ul>	<ul><li>Exchange Unified Messaging</li><li>Outlook Web App Instant Messaging and Presence</li></ul>

As shown in the preceding table, to have all the available Lync and Exchange integration features, you need to have both Lync 2013 and Exchange 2013.

At a client level, the integration between the Outlook and Lync clients differs depending on the Lync and Exchange versions. You need Lync 2013 and Exchange 2013 to be able to have the Unified Contact store, high-resolution photos (requires Lync 2013 or the Microsoft Lync Web App, for higher resolution; all other clients will downscale the pictures), and the Instant Messaging archiving. We will describe these features in detail and show you how to configure and test them later on in this chapter.

However, there are more features that you can use when integrating Outlook with the Lync client, and these features are available with Lync 2013 and both Exchange 2013 and 2010. These features, provided by the Exchange Web Services, are the following:

- Read or delete items in the conversation history folder in Outlook
- Listen to or delete voicemail items
- Display extended free busy information and meeting subject and location
- Read or delete Lync missed call notifications in your inbox

Hopefully, the preceding information gave you a good understanding of how Lync and Exchange integrate, both at the server and client levels. The new integration features provided with both Lync 2013 and Exchange 2013, although visible at the client level, need to be configured at the server level. This means that if you don't configure Lync 2013 and Exchange 2013 accordingly, you will not be using the **Unified Contact Store** (**UCS**) or archiving your Instant Messages in the Exchange mailbox.

In the upcoming sections, we will describe each of the features and provide guidance on how to configure and troubleshoot them.

Understanding Lync 2013 and Exchange Unified Messaging integration

Probably, the most well-known and used integration between Lync and Exchange is the **Unified Messaging** (**UM**) integration. To integrate Lync and Exchange at the Unified Messaging level, you need a minimum of OCS 2007 R2 and Exchange 2007. We will describe the integration between Lync 2013 and Exchange 2013 in the following sections.

Unified Messaging features

The first thing you need to know is that to be able to use the Unified Messaging feature on Microsoft Exchange 2013, you will need to have one Microsoft Exchange Enterprise CAL per user that has Unified Messaging enabled.

The following features are provided by the Exchange Unified Messaging infrastructure to Enterprise Voice enabled Lync users.

Call answering

When you don't answer a call on your Lync client, the Exchange Unified messaging will answer it and receive a voice message on behalf of the user. In essence, it's the same concept as voicemail in a traditional PBX environment. The user can record a personal greeting, and the Unified Messaging service will receive the voice message and deliver it to the user's mailbox on Exchange.

The following two things can happen in this scenario:

- The caller can leave a message that will then be delivered to the user's mailbox
- The caller hangs up the call without leaving a message, and a missed call notification e-mail is sent to the user's mailbox

Users will then be able to access their mailbox from an e-mail client such as Outlook or OWA, or from the Outlook Voice Access Unified Messaging feature, and listen to the voicemail.

Outlook Voice Access

The **Outlook Voice Access** (**OVA**) feature gives the users the ability to call the Exchange Unified Messaging service not only to listen to voicemails, but also to access their Exchange inbox, calendar, and contacts from a telephony interface.

To configure the OVA access feature, a subscriber access number is necessary; this needs to be configured in the Exchange Unified Messaging dial plan and on the Lync Server as a Unified Messaging Lync contact.

For a detailed OVA quick-start guide diagram, refer to the Microsoft Exchange TechNet article at `http://technet.microsoft.com/en-us/library/bb397228(v=exchg.141).aspx`.

Auto attendant

Auto attendant is a feature that can be configured to provide external callers with voice prompts and a navigating menu system or to allow them to forward calls to a specific number or extension. This can enable external users to reach any internal user in the company from a single external DDI associated to the auto attendant.

Some of the Auto attendant functionalities can also be achieved by configuring Lync Response Groups. By default, Auto attendant will provide you with more functionalities than the response groups.

In this chapter, we will cover all the main integration methods between Lync 2013 and Exchange 2013 to give you a better understanding of how Lync and Exchange can work together and use each other's features. We will try to be as detailed as possible on how to configure these features and give you some Lync and Exchange Management Shell commands to help you in both listing and modifying these configurations per user.

Configuring the Unified Messaging integration

Several steps must be performed to fully configure the Unified Messaging integration between Lync 2013 and Exchange 2013. We will start by describing the prerequisites to achieve this integration.

Getting ready

The following prerequisites must be in place in order to configure the Unified Messaging integration:

- **Microsoft Exchange Unified Messaging role**: The UM role should be installed on an Exchange 2007 SP1 or later, Exchange 2010 RTM or later, or Exchange 2013 RTM or later versions.
- **Certificates**: The Exchange UM certificate installed must be valid, signed by a certification authority who is trusted by the Lync Front End server(s). The Exchange server also needs to trust the certification authority who signed the Lync certificates. The Exchange certificate must have the FQDN of the Exchange Unified Messaging Server as a subject name and must ideally be issued by a private certification authority. Getting Exchange Unified Messaging from a public certification authority can be more expensive (no cost is involved when it's signed by an internal one) and even impossible if the FQDN of the Exchange Unified Messaging Server is from a non-Internet routable domain (for example, server.domain.local).
- **Enterprise Voice enabled users**: To be able to use the Unified Messaging features on the Lync client, the user must be Enterprise Voice enabled on Lync. You can use the Lync control panel to enable the user for Enterprise Voice. More details on this can be found in *Chapter 3, Lync Dial Plans and Voice Routing*.

How to do it...

Here are the steps to configure Unified Messaging integration.

Configuring the Exchange certificates for Unified Messaging integration with Lync

The first step of the configuration should be to configure valid certificates on the Exchange server and assign the Unified Messaging Call Router and the Unified Messaging service to these certificates. In Exchange 2013, the Unified Messaging Call Router service is part of the Client Access role, and the Unified Messaging Service is part of the Mailbox role. They can both coexist on the same server if you have both the Exchange 2013 roles installed on this server. The Unified Messaging Call Router service will receive all the SIP requests from Lync 2013 and redirect them to the Unified Messaging Service.

Requesting the certificate

Ideally, as I just mentioned in the previous section, you should request a certificate from an internal certification authority, for each of the Unified Messaging servers. It's cheaper and doesn't require the internal domain name to be Internet-routable. The certificate can be requested both via the Exchange Admin Centre and via the Exchange Management Shell.

To request a certificate via the Exchange 2013 EAC, proceed with the following steps:

1. Log in to the EAC as an Exchange Administrator.
2. On the left menu, click on **Servers**.
3. On the **Tab** options, click on **Certificates**.
4. Select the server you want to request the certificate to and click on **+** to request a new certificate.
5. Select **create a request for a certificate from a certification authority**.
6. Select a friendly name for the certificate.
7. Select a server to store the certificate request in.
8. On the names, make sure that you have the internal FQDN of the server.
9. Fill in the certificate details and choose the path where you want to store the request file. Click on **Finish**.

The following screenshot illustrates a certificate request done via the EAC:

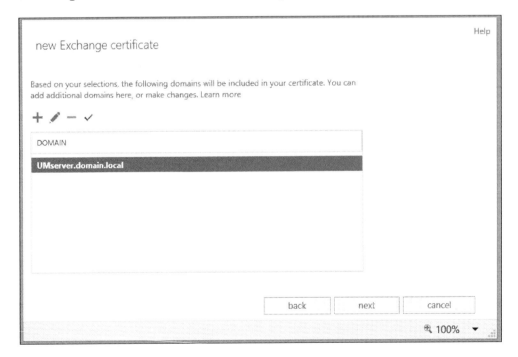

Once the certificate request is done, you should issue the certificate on your certification authority (internal or public) and then complete the request via the EAC by selecting the certificate request on the EAC and choosing the **Complete certificate request** option. You will then be prompted to enter the path for the certificate issued by your CA.

To use the Exchange Management Shell to create a new certificate, you need to run the following commands:

Command	What does the command do?
`$requestfile = New-ExchangeCertificate -GenerateRequest -SubjectName "c=US, o=Company, cn=UMserver.domain.local" -DomainName UMServer.domain.local -PrivateKeyExportable $true`	Creates a request file variable with the certificate request command
`$requestfile	out-file c:\certificaterequest.txt`	Generates the certificate request and stores the file in the specified path

The following screenshot demonstrates the commands referenced previously:

```
[PS] C:\Windows\system32>$requestfile = New-ExchangeCertificate -GenerateRequest -SubjectName "c=US, o=Company, cn=UMserver.domain.local" -DomainName UMServer.domain.local -PrivateKeyExportable $true
[PS] C:\Windows\system32>$requestfile | out-file c:\test\certificaterequest.txt
[PS] C:\Windows\system32>Get-ExchangeCertificate

Thumbprint                                Services   Subject
----------                                --------   -------
54C8E60EA3FDD25B42522B96E9A05E72AE59F146   .......   C=US, O=Company, CN=UMserver.domain.local
```

Once the certificate is requested and issued, you should complete the certificate request by importing it to the Exchange 2013 certificates computer store. The following command should be used to import the certificate via the Exchange Management Shell:

Command	What does the command do?
`Import-ExchangeCertificate -FileData ([Byte[]]$(Get-Content -Path C:\certificates\UMcertificate.crt -Encoding byte -ReadCount 0))`	Imports the generated certificate to the Exchange 2013 UM Server

For official guidance on how to request a new certificate on Exchange 2013, refer to `http://technet.microsoft.com/en-us/library/bb125165(v=exchg.150).aspx`.

Enabling the certificate

After installing the new certificate on the Exchange 2013 UM Server and making it available for the usage of Exchange Server, you need to assign the UM services to this certificate. Again, this can be done both via the EAC or the Exchange Management Shell. The following command should be used to enable the certificate via the Exchange Management Shell:

Command	What does the command do?
`Enable-ExchangeCertificate 67DA55B9F5AB2A735E7CBD11A895F7757ACC07C4 -services "UM,UMCallRouter"`	Enables the certificate with the referenced thumbprint for the Unified Messaging and Unified Messaging Call Router services

Now, let's take a look at how you should prepare your Exchange UM services to both work in TLS mode and use the certificate you just created.

To set the UM services via the EAC to work in TLS mode, proceed with the following steps:

1. Go to **servers**, and on the tab also called **servers**, double-click on your UM server to access the properties.
2. Go to the **unified messaging** option and configure both the **UM Call router settings** option and the **UM Service settings** startup mode to TLS, as shown in following screenshot. Click on **save** to exit.

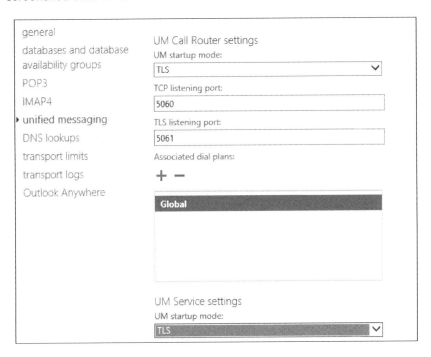

3. To set the UM services via the Exchange Management shell to work in TLS mode, run the following commands:

   `UM Service`

   `Set-UMService -Identity UMServer -UMStartupMode TLS`

   `UM Call Router Service`

   `Set-UMCallRouterSettings -Server UMServer.domain.local -UMStartupMode TLS`

Besides configuring the Unified Messaging services to run in TLS mode, you also need to enable the certificate you just created to both these services. We showed you how to do this via the Exchange Management Shell. Now, let's look at the steps to do this via the EAC.

To enable the certificate via the EAC, perform the following steps:

1. On the EAC, go to **servers** and then to the **certificates** tab. Double-click on the certificate you issued for UM. Make sure that you're not selecting the certificate used for other services.

2. Click on the **services** option and tick the checkbox for both UM services, as shown in the following screenshot. Click on **Save**.

3. As referenced previously, to enable the certificate via the Exchange Management Shell, run the following command:

   `Enable-ExchangeCertificate -Server UMServer.domain.local -Thumbprint <Certificate thumbprint> -Services UM, UMCallRouter`

4. To get the certificate thumbprint for the correct certificate, run this command:

   `Get-exchangecertificate |fl`

5. Then, check the certificate details. Take note of the thumbprint of the certificate you are going to use for UM.

> After changing the startup mode and enabling the UM certificate for both UM services, these services need to be restarted.

Creating and configuring a new Exchange UM dial plan

The next step will be to create a dial plan. This can be done, as usual, via the EAC or via the Exchange Management Shell. Once you configure the UM dial plan, you should also make sure that you configure the dialing restrictions both on the dial plan and on the UM mailbox policy.

To create a new UM dial plan via the EAC, proceed with the following steps:

1. Open the EAC and click on the **Unified Messaging** option.
2. Click on **+** to create a new dial plan. Define **Name**, **Extension length**, **Dial plan type**, **VoIP security mode**, **Audio language**, and **Country/Region code**, as shown in the following screenshot. Then, click on **Save**.

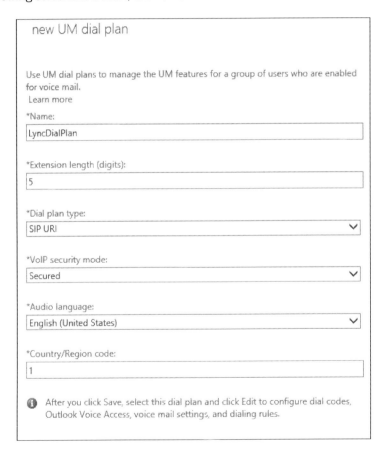

The nonoptional values when creating the dial plan are as follows:

- **Dial plan type**: Select **SIP URI**
- **VOIP security mode**: Select **Secured**

Once the dial plan is created, you should edit it, and it's fundamental that you add the following:

- An Outlook Voice Access number (for example, +44 123 456 7890).
- Dialing rules that you should create in the **Dialing rules** section and add in the **Dialing authorization** section. To have no dialing restrictions on your UM server, it's recommended that you create a rule for which the **Number Pattern** and the **Dialed Number** values are *****.

Finally, from within the UM dial plan properties, you should access the UM mailbox policy properties and also add the previously configured dialing rules on the **Dialing Authorization** section. Now, proceed with the following steps:

1. To create a new Exchange UM dial plan via the Exchange Management Shell, run the following command:

    ```
    New-UMDialPlan -Name LyncDialPlan -VoIPSecurity Secured
    -NumberOfDigitsInExtension 5 -URIType SipName -CountryOrRegionCode
    1
    ```

2. To configure the dialing restrictions and the Outlook Voice Access number for the new dial plan, run the following commands:

    ```
    Set-UMDialPlan LyncDialPlan -accesstelephonenumbers <OVA
    Number> -ConfiguredInCountryOrRegionGroups "Anywhere,*,*,*"
    -AllowedInCountryOrRegionGroups "Anywhere"
    Set-UMDialPlan LyncDialPlan -ConfiguredInternationalGroups
    "Anywhere,*,*,*" -AllowedInternationalGroups "Anywhere"
    ```

3. To configure the UM mailbox policy with the dialing rules, run the following command:

    ```
    Set-UMMailboxPolicy -Identity "<Policy Name>"
    -AllowedInCountryOrRegionGroups "Anywhere"
    -AllowedInternationalGroups "Anywhere"
    ```

    The output can be seen in the following screenshot:

4. Now that we have a new Exchange dial plan created and configured, we need to associate it to the Unified Messaging Server. To accomplish this, run the following Exchange Management Shell command:

   ```
   Set-UMService -Identity UMServer -DialPlans LyncDialPlan
   ```

5. Finally, on the Exchange 2013 Server with the Unified Messaging service and the Unified Messaging call router service, you need to make sure that both services are started. Remember that as mentioned earlier, after changing the startup mode and assigning the certificate to both services, it's mandatory to restart them.

Creating a Lync dial plan

If not created already, you should create a dial plan on Lync. The dial plan name does not need to match the Exchange UM dial plan, unless you are using Exchange 2007 SP1 or Exchange 2010 RTM.

For information on how to create a Lync dial plan, refer to the official post at `http://technet.microsoft.com/en-us/library/gg398909.aspx`.

In addition, it's recommended that you also create a normalization rule in Lync. The following command is an example of a Lync Management Shell cmdlet to create a Lync normalization rule:

```
New-CsVoiceNormalizationRule -Name 'Exchange UM' -Parent "Central"
-Description "Exchange UM" -Pattern '^(250\d{1})$' -Translation '+$1'
-IsInternalExtension: $TRUE | Out-Null
```

The preceding command assumes that you have Exchange Unified Messaging access numbers with 4 digits and starting with 250 (for example, 2501).

Running the Exchange UC Util configuration script

After creating the dial plan, we need to run the Exchange Util configuration script. This script is fundamental to get the integration with Lync working, as it will give permissions to the Lync Server on the Exchange UM objects and create a new UM IP gateway pointing to the Lync pool. You can run the script as many times as you like, and you should make sure that at the bottom of the output of the script, the dial plan you created and the Lync pool FQDN are present.

To run the script, open the Exchange Management Shell and navigate to the `scripts` directory under the Exchange install path. As an example, navigate to `C:\Program Files\Microsoft\Exchange Server\v15\scripts`.

Then, run the following command:

```
.\ExchUCUtil.ps1
```

You can see the initial output of the script as follows:

```
[PS] C:\Program Files\Microsoft\Exchange Server\V15\Scripts>.\ExchUCUtil.ps1
Using Global Catalog: GC://DC=demo-suite,DC=local
Configuring permissions for demo-suite.local\RTCUniversalServerAdmins ...
```

Chapter 4

 You might find it necessary to run the script twice before it picks up all the UM information. Once the script finishes, make sure that at the bottom of the output you have the **UMIMGGateway** and**DialPlans** columns filled with the correct information. If not, rerun the script.

Running the Lync Unified Messaging configuration tool

Now, we need to create Lync contacts for UM for the subscriber access number and auto attendants, if you have any. To do this, navigate to `C:\Program Files\Common Files\Microsoft Lync Server 2013\Support`.

Then, run the `OcsUmUtil.exe` tool.

From within the tool, proceed with the following steps:

1. Click on **load data** and then on **add** to add a new subscriber access contact.
2. Enter all the details for the contact, including whether it's subscriber access or auto attendant and the number associated with it.

Once the preceding steps have been completed, close the tool. The following screenshot shows the options present in the tool:

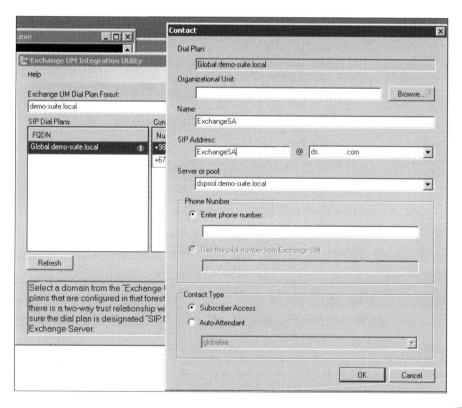

Lync 2013 Integration with Exchange

> Once the preceding steps have been executed, force an address book on the Lync 2013 side by running the following cmdlets on the Lync 2013 Shell:
> `Update-CsUserDatabase`
> `Update-CsAddressBook`

You can check the contacts created by the tool via the Lync Management Shell by running the following command:

`Get-CsExUmContact |fl`

The output of the Shell can be seen in the following screenshot:

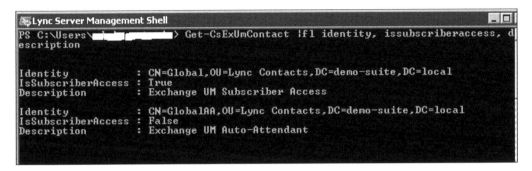

Finally, we need to restart the Lync 2013 Front End service.

Once this is done, the integration between Lync 2013 and Exchange 2013 at the Unified Messaging level is done!

Managing your Unified Messaging users

Although the task of managing your Unified Messaging users is exclusive to Exchange Administrators, we decided to include a small section on the subject. We will show you how to enable a user for UM, change the user UM settings, or get a list of all the UM-enabled users and their main settings.

Enabling users for Unified Messaging

You can use both the Exchange Administrative Centre and the Exchange Management Shell to enable your Exchange users for UM.

To enable a user for UM via the EAC, perform the following steps:

1. Log in to the EAC as an Exchange Administrator.
2. Go to **Recipients** and select your user. On the right menu, under **Phone and Voice Features**, click to enable Unified Messaging.

Chapter 4

3. Select **Unified Messaging Mailbox Policy** and click on **Next**.
4. If the user is enabled for Lync and defined as an extension, those values should be prepopulated, and you just need to click on **Finish**, as shown in the following screenshot:

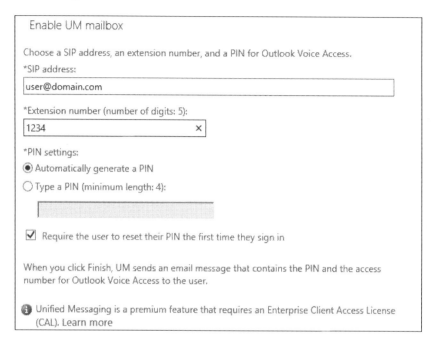

5. To enable a user for Unified Messaging via the Exchange Management Shell, run the following command:

    `Enable-UMMailbox -Identity user@domain.com -UMMailboxPolicy UMMailboxPolicy -Extensions 1234`

6. Once the user is enabled for Unified Messaging, they will receive an e-mail with all the Exchange 2013 Unified Messaging details for their account.

Changing the Unified Messaging settings for a user

You can access the user's Unified Messaging details using the EAC or running the Exchange Management Shell cmdlet, `Get-UMMailbox`. You can change these settings via the EAC by going to the same section where you went to enable the user and clicking on **view settings**, or you can do it using the `Set-UMMailbox` cmdlet. A lot of settings can be changed on the user UM properties, such as the phone number. You can also enable or disable the Play on Phone feature or the subscriber access. Refer to the official link at http://technet.microsoft.com/en-us/library/bb124893(v=exchg.150).aspx for guidance on how to change the UM settings and a list of what settings can be changed.

Lync 2013 Integration with Exchange

There's more...

In addition to giving you the recipes to configure the Unified Messaging integration between Lync 2013 and Exchange 2013, you can find information in the following sections on how to list Unified Messaging attributes, export this information, and test the Unified Messaging configuration, among others. Enjoy!

Using the Exchange Management Shell to export users' Unified Messaging details

The Exchange Management Shell is a powerful tool that you can use to get information such as:

- List of users with Exchange UM enabled and disabled
- List with all the display names, SIP addresses, and extensions for UM-enabled users
- Number of UM-enabled users per mailbox policy

The preceding examples are just a few among many features that you can use the Exchange Management Shell to achieve.

Now, let's see how we can achieve this information by running an Exchange Management Shell cmdlet.

List of users with Exchange UM enabled and disabled

To list all the users with Unified Messaging enabled, run the following command:

```
get-mailbox |where-object {$_.UMEnabled -eq $true}
```

To list all users with unified messaging disabled, run the following command:

```
get-mailbox |where-object {$_.UMEnabled -eq $false}
```

To export this list into a CSV file, with the name, alias, and UMEnabled status, run the following command:

```
get-mailbox |where-object {$_.UMEnabled -eq $false} |Select-Object Name, Alias, UMEnabled |export-csv c:\test\umdisabledusers.csv -NoTypeInformation
```

List of UM-enabled users, with display names, UM mailbox policy, and extensions

To export a list of the display names, UM mailbox policy, and extensions of all your Unified Messaging users to CSV run the following command:

```
Get-UMMailbox |select-object DisplayName, UMMailboxPolicy, @{Name='Extensions';Expression={[string]::join(";",($_.Extensions))}} |Export-Csv c:\test\umusers2.csv -NoTypeInformation
```

Number of UM-enabled users per mailbox policy

To get a list of users using a specific UM mailbox policy, run the following command:

```
Get-UMMailbox |where-object {$_.UMMailboxPolicy -eq "Global Default
Policy"}
```

To count the number of users using a specific mailbox policy, run the following command:

```
(Get-UMMailbox |where-object {$_.UMMailboxPolicy -eq "Global Default
Policy"}).count
```

Test the Unified Messaging connectivity

To test the UM connectivity, run the following cmdlet from the Lync Management Shell:

```
$credential = Get-Credential "domain.local\senderusername"
Test-CsExUMVoiceMail -TargetFqdn "lyncpool.domain.com"
-ReceiverSipAddress "sip:receiversipaddress@domain.com" -SenderSipAddress
"sip:sendersipaddress@domain.com" -SenderCredential $credential
```

The output of the preceding PowerShell cmdlet should be **Success**. Then, check the receiver mailbox and see whether the test command sent a test voicemail to their inbox successfully.

See also

For some other tips on the Exchange Unified Messaging configuration and specific scenarios such as multilanguage auto attendants, refer to the excellent blog article at http://msucblog.wordpress.com/2013/04/09/multilanguage-auto-attendant-with-ms-exchange-2010-um-and-lync/.

Configuring OAuth between Lync 2013 and Exchange 2013

Now you will learn how to configure OAuth server-to-server authentication between Lync 2013 and Exchange 2013 and also learn how to configure them both to be partner applications of each other.

This step is a prerequisite when you wish to achieve the following levels of integration:

- Lync 2013 to use Exchange 2013 for archiving
- Lync 2013 to use the Exchange 2013 unified contact store
- Lync 2013 to use high-resolution photos on Exchange 2013

To configure OAuth, we will assign server-to-server authentication certificates on Lync and Exchange to allow them to communicate with each other.

Lync 2013 Integration with Exchange

So, what certificates shall we use? For Exchange, we can use the default self-signed Microsoft Exchange Server Auth Certificate that is created when you install your Exchange Server. For Lync, you can use your existing Lync certificate for server-to-server authentication, provided that:

- The certificate has the SIP domain on the subject field
- The same certificate is used as OAuthTokenIssuer on all frontend servers
- The certificate length is at least 2048 bits

For more information on how to assign a server-to-server authentication certificate on Lync 2013, refer to `http://technet.microsoft.com/en-us/library/jj205253.aspx`.

How to do it...

Now that we've covered the certificates, the first thing you need to do is make sure that your Exchange 2013 autodiscover service is fully functional.

To verify your autodiscover settings, run the following command:

`Get-ClientAccessServer | ft Name, AutoDiscoverServiceInternalUri`

This will provide you with a list of the `AutoDiscoverServiceInternalUri` values for each client access server on your Exchange organization. But what is `AutodiscoverServiceInternalUri`?

Well, it's the **service connection point** (**SCP**) that domain-joined machines on the internal network will use to reach Exchange and make an autodiscover query.

The following value is the recommended value for the SCP. Make sure that the FQDN is a subject alternative name on the certificate used by IIS on the Exchange Server (for example, `autodiscover.domain.com`):

`https://autodiscover.domain.com/autodiscover/autodiscover.xml`

The following value is the default value for the SCP on a newly installed Exchange 2013 server, which is the FQDN of the server:

`https://exchangeserver.domain.local/autodiscover/autodiscover.xml`

If the value is blank or if the value points to a name that is not on the certificate that your Exchange Server uses for IIS, you should change it. If you are load balancing the Exchange Web Services between more than one Exchange 2013 Client Access Server, make sure that you assign the same name to all of the `autodiscoverserviceinternaluri` values on each of these servers (that is, `autodiscover.domain.com`) and again make sure that the name is on the certificate used by Exchange for IIS. Point the `autodiscover.domain.com` DNS record to the IP address of the load balancer configured to balance the Exchange 2013 HTTPS traffic.

To change the value or configure one, you can run the following Exchange Management Shell cmdlet:

```
Get-ClientAccessServer | Set-ClientAccessServer
-AutoDiscoverServiceInternalUri https://autodiscover.domain.com/
autodiscover/autodiscover.xml
```

The preceding command will change the attribute on all the Client Access servers of your organization.

The following command changes the value on a specific client access server:

```
Set-ClientAccessServer -Identity <CASServerName>
-AutoDiscoverServiceInternalUri https://autodiscover.domain.com/
autodiscover/autodiscover.xml
```

The last step to complete the OAuth configuration is to modify the Lync Server OAuth configuration settings to make sure that Lync can find the Exchange autodiscover service. To achieve this, we need to run the following Lync Server Management Shell cmdlet:

```
Set-CsOAuthConfiguration -Identity global -ExchangeAutodiscoverUrl
https://autodiscover.domain.com/autodiscover/autodiscover.svc
```

> The URL configured on Lync should point to the service location (autodiscover.svc) and not to the XML file (autodiscover.xml) used by the autodiscover service. You can test whether the URL works using a web browser.

Once again make sure that the DNS name you use on the URL points to the Exchange Server or to a load balancer that is balancing the web traffic between your Exchange Client Access Servers.

Configuring Lync 2013 and Exchange 2013 as partner applications

After configuring OAuth and the certificates, we must now instruct Lync 2013 and Exchange 2013 to be partner applications. This will allow Lync 2013 and Exchange 2013 to communicate and interoperate securely.

How to do it...

Here is how we configure Lync 2013 and Exchange 2013 as partner applications.

Configuring Lync 2013 to be a partner application on Exchange 2013

Exchange 2013 comes with a script to quickly do this configuration. The only thing you need to know and understand is which URL you will have to configure when running the script on Exchange. The URL should have the FQDN of your Lync pool and look like `https://lyncpool.domain.com/metadata/json/1`.

> The Lync pool name must be on the web certificate used on the Lync Front End server. If the Lync pool is a standard pool, the name of the pool is the server FQDN.

So now, let's explain how to run the script. First, open the Exchange Management Shell and browse to the `scripts` directory under the Exchange installation path (that is, `C:\Program Files\Microsoft\Exchange Server\V15\Scripts\`). Then, run the following script:

```
.\Configure-EnterprisePartnerApplication.ps1 -AuthMetaDataUrl https://lyncpool.domain.com/metadata/json/1 -ApplicationType Lync
```

After running the script, you need to perform an IIS reset on your Exchange Client Access servers.

Configuring Exchange 2013 to be a partner application on Lync 2013

Now, we need to configure Lync 2013 to have Exchange 2013 as a partner application. In this case, we also need to know which URL we will include on the Lync cmdlet that will configure the integration. The URL should point to the `autodiscover` URI configured on the Exchange Client Access servers and look like `https://autodiscover.domain.com/autodiscover/metadata/json/1`.

Now that we have explained the URL, let's run the command.

Open the Lync Management Shell and run the following cmdlet:

```
New-CsPartnerApplication -Identity Exchange -ApplicationTrustLevel Full -MetadataUrl https://autodiscover.domain.com/autodiscover/metadata/json/1
```

We have now configured both platforms as partner applications.

Test OAuth and the partner applications

Once all the previous configurations are completed, we can use a test cmdlet provided by Lync to make sure that the integration is working.

The test cmdlet that we will use is the following:

```
Test-CsExStorageConnectivity -SipUri sip:user@domain.com
```

The SIPUri should be the SIP address of the user with a mailbox in Exchange 2013. The preceding cmdlet will try and write on the conversation history folder of the user's mailbox, and if the command runs successfully, it indicates that you have both OAuth and the partner applications well configured.

Configuring Lync 2013 to use Exchange 2013 for archiving

With Lync 2013 and Exchange 2013, you can choose to archive Instant Messaging and web conferencing transcripts into the user's mailbox rather than doing it into a SQL database configured for the Lync archiving role.

The data will be archived in a hidden folder called `Purges` within the user's mailbox. The `Purges` folder is under the recoverable items folder in the mailbox of the user. You could then perform an eDiscovery search in the user's mailbox to find those items, if required. You need to be aware that the Lync data archived requires storage on Exchange and, therefore, needs to be taken into account when designing the Exchange 2013 Servers, especially when calculating the storage space. If the Exchange 2013 infrastructure is already in place, you need to make sure that the disk space on the Exchange 2013 Servers allows this feature to be implemented. For more information on how to perform an Exchange eDiscovery search, refer to `http://technet.microsoft.com/en-us/library/dd353189(v=exchg.150).aspx`.

How to do it...

There are three steps that you need to complete, all on Lync, to get archiving on Exchange working:

1. Enable the Lync archiving configuration.
2. Enable archiving for internal or external communications.
3. Configure the `ExchangeArchivingPolicy` property per user (not mandatory).

Enable Exchange archiving

On Lync, by default, both archiving and Exchange archiving are not enabled. To be able to archive on Exchange, you need to edit the global archiving configuration or create a new archiving configuration. To enable archiving on Exchange on the global archiving configuration, you need to run the following cmdlet on the Lync Management Shell:

```
Set-CsArchivingConfiguration -Identity "global" -EnableArchiving ImOnly
-EnableExchangeArchiving $True
```

> When you enable archiving, you can define it to be **None**, **IMOnly** (archives only Instant Messaging), or **IMandWebConf** (archives Instant Messaging and web conferencing sessions).

Enable archiving to internal or external communications

Once you've configured the Lync archiving configuration, you must now edit the global archiving policy to define which type of communications will be archived: internal or external. It is recommended that if you have sets of users with different archiving configurations to be applied, you could create several archiving policies and configure them accordingly.

Usually, both the legal and the human resources departments should be involved in the discussion of what Microsoft Lync 2013 data to archive and for which users.

To edit the global archiving policy, run the following cmdlet:

```
Set-CsArchivingPolicy -Identity "global" -ArchiveInternal $True
-ArchiveExternal $True
```

To create a new archiving policy, run the following command:

```
New-CsArchivingPolicy -Identity "NewPolicyName" -ArchiveInternal $True
-ArchiveExternal $True
```

Configuring the ExchangeArchivingPolicy property per user

You can configure, per user, to not archive. Use Lync for archiving or use Exchange. To configure a user to archive on Exchange, run the following command from the Lync Management Shell:

```
Set-CsUser -Identity "User" -ExchangeArchivingPolicy ArchivingToExchange
```

Making sure that Lync is archiving on the user Exchange mailbox

The fastest and most reliable way to make sure that Lync is archiving on the user mailbox in Exchange is to use the MFCMAPI tool to look at the `Purges` folder under the recoverable items hidden folder in the user mailbox. You should use this tool with caution and, ideally, on a test mailbox enabled for archiving to prevent accidental deletion of data on a production mailbox. Now, proceed with the following steps:

1. To download the MFCMAPI tool, refer to `http://mfcmapi.codeplex.com/`.

2. To check the `Purges` folder, open MFCMAPI, click on **Session\Logon** and choose the Outlook profile you want to open, double-click on your mailbox, authenticate it if needed, and then browse to the `purges` folder, which is shown in the following screenshot:

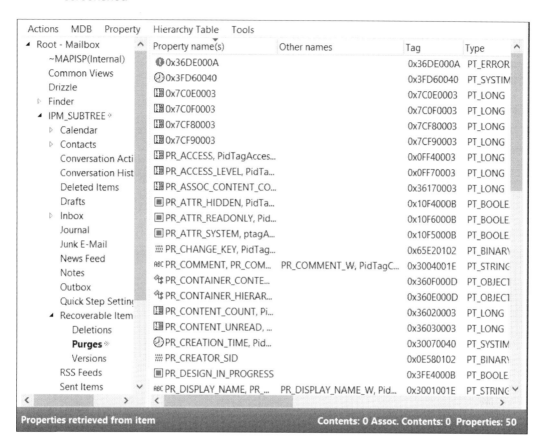

3. Double-click on the `purges` folder and look for the archived Instant Messages from Lync.

There's more...

Now that the archiving is configured, use the following recipes to get and export some useful information on archiving to your organization.

Use the Lync Management Shell to get reports for your users' archiving configurations

Via the Lync Management Shell, there are a number of reports on the users who archive configuration; you can obtain these reports. Now, we will provide you with Lync Management Shell cmdlets to get some of these reports.

List Lync users with the Exchange archiving policy property defined to archive in Exchange or Lync

To obtain a list of Lync users with the Exchange archiving policy set to archive on Exchange, run the following cmdlet:

```
Get-Csuser |Where-Object {$_.ExchangeArchivingPolicy -eq "ArchivingtoExchange"}
```

To obtain a list of Lync users with the Exchange archiving policy set to archive on Lync, run the following cmdlet:

```
Get-Csuser |Where-Object {$_.ExchangeArchivingPolicy -eq "UseLyncArchivingPolicy"}
```

List or change the user's Lync archiving policy

To get a list of your Lync users who have assigned a certain Lync archiving policy, run the following cmdlet:

```
Get-CsUser |Where-Object {$_.ArchivingPolicy -eq "LyncArchivingPolicy"}
```

To assign an archiving policy to a user or set of users, run the following cmdlet:

```
Get-CSUser | Grant-CsArchivingPolicy -PolicyName "LyncArchivingPolicy"
```

Configuring Lync 2013 to use the Exchange 2013 Unified Contact Store

The **Unified Contact Store** (**UCS**) enables you to store all the user's contacts on Exchange. The user will store all their contacts in a single location and retrieve them via the Exchange Web Services.

Getting ready

With the server-to-server authentication configured, the only thing you need to do is configure the Lync user services policy to allow access to the Unified Contact store, which is not enabled by default, to a user or set of users. It's recommended that if you want to have different UCS configurations from different groups of users, you could create one or several new user services policies.

To change the global user services policy to use UCS, run the following Lync Management Shell cmdlet:

```
Set-CsUserServicesPolicy -Identity global -UcsAllowed $True
```

Note that the default value of the `UcsAllowed` attribute is `true`, and therefore, you don't need to change it if the value is still set to `default`.

To create a new user services policy that does not use UCS, run the following cmdlet:

```
New-CsUserServicesPolicy -Identity "DenyUCS" -UcsAllowed $False
```

How to do it...

Let's see how to configure Lync 2013 to use the Exchange 2013 UCS in the following sections.

Manage and list your UCS settings

You can use the Lync Management Shell to either get reports of the UCS settings per user or to change them.

List all the users with UCS enabled

To get a list of all the users who use UCS, run the following cmdlet to get a list of the user services policies that allow UCS:

```
Get-CsUserServicesPolicy |where-object {$_.UCSAllowed -eq $true}
```

Then, get a list of users that use that user services policy:

```
get-csuser |where-object {$_.UserServicesPolicy -eq "Global"}
```

Change the user services policy for a user or set of users

To change the user services policy for a user, run the following cmdlet:

```
Grant-CsUserServicesPolicy -Identity "User" -PolicyName "DenyUCS"
```

The preceding command will apply a policy that does not use UCS and, therefore, disable that feature for the specified user.

Test the Unified Contact Store feature

There are two ways of testing the UCS feature for a specific user.

The first method is to run the following Lync Management Shell cmdlet:

```
Test-CsUnifiedContactStore -UserSipAddress "sip:user@domain.com"
-TargetFqdn "lyncpool.domain.com"
```

The user SIP address for a user with an Exchange mailbox and the Lync pool FQDN should be specified in the preceding command.

The second way of making sure that the UCS is working for a user is by checking the UCS settings on their Lync client configuration information.

To do this, hold the *Ctrl* key and right-click on the Lync client icon on the system tray of a client computer. Then, choose **Configuration information**.

From the configuration information, check the **UCS Connectivity State** option to see whether UCS is enabled and being used.

Integrating Lync 2013 with the Exchange 2013 Outlook Web App

With Lync and Exchange 2010 or 2013, you can configure both platforms so that users can use the Exchange Outlook Web App for Instant Messaging and presence.

Getting ready

The first thing you need to check to make sure that you can do the integration is whether **Unified Communications Managed API 4.0 Runtime** is installed on your mailbox servers. To verify this, log in to your mailbox servers and look for the following registry key:

```
HKEY_LOCAL_MACHINE\SYSTEM\CurrentControlSet\Services\MSExchange OWA\
InstantMessaging\ImplementationDLLPath
```

This key should point to the folder location of the `Microsoft.Rtc.Internal.Ucweb.dll` file. If the key does not exist or does not point to this folder, you should install the application on all of your mailbox servers.

You can also check the installed programs via the control panel of your mailbox servers.

Refer to `http://www.microsoft.com/en-us/download/details.aspx?id=34992` to download and install the **Unified Communications Managed API** (**UCMA**).

How to do it...

To configure the Lync 2013 and Exchange 2013 OWA integration, you need to follow several steps in the same order as described in the following sections.

Creating a trusted application pool on Lync for OWA

The next step is to create a trusted application pool on Lync for the Outlook Web App.

> If you already have Unified Messaging integration between Lync and Exchange and if your Exchange 2013 servers have both the Client Access and the Mailbox roles, you *do not* need to create the trusted application pool on Lync. Creating a trusted application pool in this scenario will break the Instant Messaging and presence integration with OWA.

To create the trusted application pool, run the following cmdlet:

```
New-CsTrustedApplicationPool -Identity <ExchangeServerFQDN> -Registrar lyncpool.domain.com -Site SiteName -RequiresReplication $False
```

The identity of the trusted application pool should be the FQDN of the Exchange Client Access server or the name that is being load balanced to reach the Client Access server (that is, `owa.domain.com`).

If you want to get the SiteName, run the following cmdlet:

```
Get-CsSite |ft displayname, siteID
```

The register value should be the FQDN of the Lync pool.

Once the trusted application pool is created, you need to create the trusted application by running the following Lync Management Shell cmdlet:

```
New-CsTrustedApplication -ApplicationId OutlookWebApp -TrustedApplicationPoolFqdn <ExchangeServerFQDN> -Port 5199
```

Once both are created, enable the Lync topology by running this command:

```
Enable-CsTopology
```

Configuring Exchange for the IM integration with OWA

There are three things you need to do on Exchange to complete the Instant Messaging integration with OWA:

1. Configure the OWA virtual directories for Instant Messaging.
2. Edit the `web.config` file on the Client Access servers.
3. Configure the OWA mailbox policy to allow Instant Messaging on OWA.

Configuring the OWA virtual directories for Instant Messaging

You need to enable on one or more OWA virtual directories on your Exchange Client Access servers for Instant Messaging integration. If there are some Client Access servers used by some users within your organization and you don't wish these client access servers to provide the IM on the OWA functionality, then you should just enable the properties on the other servers.

To get a list of all the OWA virtual directories and the Instant Messaging current configurations, run the following Exchange Management Shell cmdlet:

```
Get-OwaVirtualdirectory |ft identity, instantmessagingenabled, instantmessagingtype
```

To enable Instant Messaging on a specific virtual directory of a specific server, run the following cmdlet:

```
Set-OwaVirtualDirectory -identity <OwaVDirIdentity> -instantmessagingenabled $true -instantmessagingtype OCS
```

To enable Instant Messaging on all of your Client Access servers, run the following cmdlet:

```
Get-OwaVirtualDirectory | Set-OwaVirtualDirectory -instantmessagingenabled $true -instantmessagingtype OCS
```

Editing the web.config file on your client access servers

To edit the `web.config` file, proceed with the following steps:

1. On each of your Client Access servers, edit the `web.config` file present at `C:\Program Files\Microsoft\Exchange Server\V15\ClientAccess\Owa` (depends on the Exchange installation path).

2. Before editing the file, you need to know the thumbprint of the certificate that your Client Access server uses. To get this information on the Client Access server Exchange Management Shell, run the following cmdlet:

   ```
   Get-ExchangeCertificate | Fl
   ```

3. Make sure that you write down the certificate thumbprint. The certificate used must have the FQDN of the server on the subject name and as an alternative name.

4. After getting this information, you need to open the `web.config` file and edit the following section:

   ```
   <add key="IMCertificateThumbprint" value="<Certificate Thumbprint>"/>
   
   <add key="IMServerName" value="lyncpool.domain.com"/>
   ```

The explaination for the preceding command can be explained as follows:

- In `IMCertificateThumbprint`, you should copy and paste the thumbprint you obtained previously
- In `IMServerName`, you should enter the FQDN of the Lync Server pool

 It's recommended that you back up the `web.config` file before editing it.

5. Finally, you need to recycle the application pool on the Client Access server by running the following cmdlet:

   `C:\Windows\System32\Inetsrv\Appcmd.exe recycle apppool /apppool.name:"MSExchangeOWAAppPool"`

Configuring the Outlook Web App mailbox policy

Each user has an Outlook Web App mailbox policy assigned, and by default, the OWA mailbox policy will not allow Instant Messaging on OWA. You can either configure the default policy to allow all users to have this feature enabled, or you can create new policies and assign them to the relevant users.

To modify the default mailbox policy to allow Instant Messaging on OWA, run the following Exchange Management Shell cmdlet:

`Set-OwaMailboxPolicy -Identity "Default" -InstantMessagingEnabled $True -InstantMessagingType "OCS"`

To configure all of your existing mailbox policies to allow Instant Messaging on OWA, run the following Exchange Management Shell cmdlet:

`Get-OwaMailboxPolicy | Set-OwaMailboxPolicy -InstantMessagingEnabled $True -InstantMessagingType "OCS"`

To get a list of all your mailbox users that have a specific mailbox policy, run the following cmdlet:

`Get-CASMailbox |where-object {$_.OwaMailboxPolicy -eq "Default Policy"}`

5
Scripts and Tools for Lync

In this chapter, we will cover the following recipes:

- Installing Lync prerequisites and more – Set-Cs2013Features
- Creating a fully functional voice configuration – Lync Dialing Rule Optimizer
- Switching between multiple Lync identities with a click – Profiles for Lync (P4L)
- Tracing made easier – Lync 2013 Centralized Logging Tool
- Identifying recurrent issues – Lync Pilot Deployment Health Analysis
- Managing phone numbers – Search-LineURI and Get-UnusedNumbers
- Managing call pickup groups – Lync2013CallPickupManager1.01

Introduction

Lync Server 2013 is indisputably a well-engineered product. Built on the foundations of **PowerShell** engine, it also features administrative capabilities through a web-based control panel. Consistent with Microsoft's practice, only around 80 percent of PowerShell capabilities are exposed and exploitable through the administrative GUI; that is, PowerShell is the only option for advanced or automated tasks. Regardless of this, the Lync control panel does quite well in providing the majority of the most common day-to-day administrative tasks in an easy and intuitive way. This is quite a welcome relief for the less experienced Lync administrator!

That said, Lync still has room for improvements, with a few areas where manageability is not as straightforward as it could be. Furthermore, PowerShell is a complex subject that requires a considerable investment of time before it can be mastered.

Scripts and Tools for Lync

Luckily, there are many solutions that will make life easier in certain areas where out-of-the-box features fall a bit short, providing quick and intuitive ways to perform specific tasks. Some are commercial (paid) add-ons, while others are free software. Our aim is to focus on the latter. The tools presented here have the following common features:

- They are free and often open source. Commercial (paid) tools and add-ons are not included deliberately.
- They are written by well-known, knowledgeable community members who love to share their great skills and passion for Lync.

Be advised that these tools are unofficial and unsupported software provided as is and without warranties of any kind from this book's author or the software's author. Use them at your own risk.

In this chapter, we will present a number of recipes to help speed up typical Lync deployment, configuration, and administration tasks through a selection of the best tools available. You will find more special-purpose tools (like the ones for backup, planning, and troubleshooting) in other chapters of this book.

For more information about Lync PowerShell cmdlets, take a look at http://technet.microsoft.com/en-us/library/gg398867.aspx.

Installing Lync prerequisites and more – Set-Cs2013Features

Like many other Microsoft server products, Lync 2013 requires some software prerequisites to be available before installing Lync Server roles. If a prerequisite is missing, you will be warned, and installation will fail. Types of prerequisites range from the operating system level (that is, IIS components) to the application level; a few others are at the component level.

There is no built-in prerequisite installation wizard in Lync; before installing it, you will need to be aware of the exact components to preinstall (through the GUI, PowerShell, or a combination of both). Further to that, each Lync Server role has different software prerequisites.

A number of scripts were put together by Lync enthusiasts to overcome this. The most notable one to date is Set-CsLync2013Prerequisites by Microsoft Lync MVP, Pat Richard (@patrichard). The first release dates back to Lync 2010; it's therefore a mature and stable tool, and is actively updated by the author. Originally aimed at automating the prerequisites' installation, it has now evolved into a nifty Swiss Army knife multipurpose utility (over 40 tools at the time of writing), which also take care of automating several other recurrent administrative or post-installation tasks.

Getting ready...

The tool consists of a PowerShell script (.ps1), and it is available on the author's blog at http://www.ehloworld.com/1697.

You will need to perform the following tasks in order to run it:

- Install the **Lync 2013 administrative tools** from the Lync Server installation media. This will add PowerShell modules for Lync. Ensure that you are a local administrator. You will also need to be a member of the CsAdministrator group for some functions. This group is created as part of the AD DS schema and forest preparation for Lync.

- Ensure that you set up PowerShell to run signed external scripts. From an elevated (administrative) PowerShell, run Set-ExecutionPolicy AllSigned. For more information about the implications of allowing scripts to run, take a look at http://technet.microsoft.com/en-us/library/ee176961.aspx.

- Ensure that the server has Internet access. Several components will be automatically downloaded.

> Proxy support is still somewhat unreliable. If you are behind a proxy server, the download might fail.

How to do it...

To run the script, type the following:

Command	What does the command do?
.\Set-Cs2013Features.ps1	This runs the script with default values
.\Set-Cs2013Features.ps1 -Win2012Source "d:"	This runs the script and sets the Windows 2012 installation media source. The script will require access to it to install some software.
.\Set-Cs2013Features.ps1 -SQLPath "d:\sqlexpress"	This runs the script and sets a custom SQL Server Express installation folder (if not specified, it defaults to C:\Program Files\Microsoft SQL Server).

When running the tool, you will be presented with a list of possible tasks, specifically to install the prerequisites, as shown in the following screenshot:

```
1) Director prerequisites
2) Edge prerequisites
3) Front-End prerequisites
4) Mediation prerequisites
5) Office Web Apps Server 2013 (prereqs, OWAS, SP1, EN lang pack)
6) Persistent Chat prerequisites
```

Options 1 to 6, will install Lync prerequisites. You will need to know what Lync role you are installing and ensure that you select the correct one. If you do not, you might end up missing components for the role or install unnecessary ones. One thing to notice is that all Lync 2013 server roles require a **SQL Server Express** installation. Unless already installed, the Lync setup wizard will take care of this. Lync 2013 does not have an automated prerequisite installation capability. The exception is the automatic installation of SQL Server Express RTM. However, if you let Lync wizard do the job, an outdated SQL Express RTM version is installed. The `Set-Cs2013Features` script goes a step further and will install SQL Server Express along with service packs and updates released at the time of installation.

There's more...

Office Web Apps is an Office server that is required for the purpose of presenting PowerPoint presentations in meetings. This is a new requirement in Lync 2013, and this server cannot be collocated with any other Lync Server. The prerequisite tool does a great job by automating the following tasks:

- Downloads and installs Office Web Apps server components from the Internet. Note that downloads are quite large.
- Downloads and installs Office Web Apps service packs and other updates as required.

Options 7 to 12 will automatically download and install the components shown in the following screenshot:

```
7) Lync Server 2013 Resource Kit
8) Lync Server 2013 Persistent Chat Resource Kit
9) Lync Server 2013 Debugging Tools
10) Lync Server 2013 Stress and Performance Tool
11) Lync Server 2013 Best Practices Analyzer
12) Lync Server Connectivity Analyzer
```

Several other self-explanatory features are available in the tool, ranging from installation of other tools such as Wireshark or Microsoft Message Analyzer, to one-stop features to fix common issues, as shown in the following screenshot:

Chapter 5

```
15) Launch Windows Update
16) SCOM Watcher Node prerequisites
17) Custom PortQryUI
18) Microsoft Message Analyzer (formerly NetMon)
19) Add custom Scheduler URL to simple URLs
20) SQL Server Management Studio
24) Microsoft UCMA 4.0 - for sefautil.exe
28) Configure Skype Federation
30) WireShark
31) Enable Photo URL option
34) Lync Room System Admin Portal prerequisites
36) Create Lync file share on local computer

50) Misc server config menu  -->
60) Desktop shortcuts menu   -->
70) Taskbar shortcuts menu   -->
80) Downloads only menu      -->
90) Security menu            -->

97) Visit website for this script
98) Restart the server
99) Exit
```

See also

- For more information about software prerequisites for Lync 2013 Server installation, refer to http://technet.microsoft.com/en-gb/library/gg398103.aspx.

Creating a fully functional voice configuration – Lync Dialing Rule Optimizer

Microsoft Lync with Enterprise Voice provides a full standalone VoIP solution that can replace the traditional PBX or enhance it. However, voice configuration in Lync is neither simple nor exactly intuitive as several elements such as Dial Plans, PSTN usages, Voice Policies, and Routes are involved. The availability of public documentation on the subject is large, however, often resulting in a challenging and confusing experience for the less experienced user.

Enterprise Voice is Microsoft's wording for Lync interoperability with PSTN voice.

Specifically to dial plans, routes, and trunk translation rules, you will also need good knowledge of **Regular Expressions** (**RegEx**). This is the adopted methodology to match specific strings of characters, with the purpose of performing E.164 phone number normalization/manipulation or determining the call route based on the called number (references to RegEx are provided in the *See also* section).

Scripts and Tools for Lync

> **E.164** is an international numbering standard for public telephone systems in which each number must contain a country code, a national access code, and a subscriber number, therefore resulting in a unique and unambiguous identification worldwide. You might regard E.164 as the rough equivalent of a fully qualified domain name for DNS. An example of E.164 number is +44201234567 (44 is the country code for United Kingdom, 20 is the area code for London, and 1234567 is the subscriber number).

The **Lync Dialing Rule Optimizer** is a web-based tool by Lync MVP, Ken Lasko (`@kenlasko`). It is simple and intuitive to use, and is designed to take care of most typical Enterprise Voice setups. Despite the word "optimizer" in the name, the tool is actually capable of creating a fully-functional voice configuration from scratch.

Getting ready...

The tool is available at `http://www.lyncoptimizer.com/`. You will need to sign in through a Microsoft Live account; this will allow you to retrieve your usage history, as shown in the following screenshot:

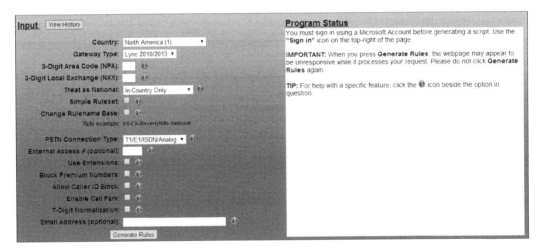

Chapter 5

How to do it...

Let's walk through the main parameters you will need to specify, with a practical example. For this purpose, we will assume you want to configure a direct SIP trunk for an office based in New York City, and you got the following continuous numbering range assigned from your PSTN carrier:

- The first number is 212-555-1001 (in E.164: +12125551001)
- The last number is 212-555-1099 (in E.164: +12125551099)

 We assume that your SIP trunk is already configured as a PSTN gateway in the Lync topology. Lync Optimizer requires this before being able to run the output script.

The following are the main parameters we will use:

- **Country**: At the time of writing, the tool can create dialing rules for over 75 countries, and the list keeps growing. For the purpose of our walkthrough, select North America (the country code is 1).

- **Gateway type**: Select Lync 2010/2013.

- **3-Digit Area Code**: Enter the area code where your Lync Server or PSTN gateway resides. In our example, enter 212 for New York.

- **3-Digit Local Exchange**: These are the first 3 digits in the subscriber number range that you are assigned, and it usually corresponds to a specific geographic area. In our example, enter 555.

- **Simple ruleset**: If you select this option, the tool will create a single PSTN usage named *AllCalls* instead of multiple usages that separates local, national, mobile, premium, toll free, service, and international call types. This option should only be used if you want to allow users to call any number without restrictions. Typically, you will want to assign specific calling privileges (also called class of usage) to different users.

- **Change rulename base**: Use this if you want to change the default naming scheme used by the Optimizer, and type a name base that is easier for you to remember. Check this box and type `NewYork`.

- **External Access #**: This is a legacy PBX feature, where you needed to dial a prefix to get access to the outside line (typically 0 or 9). You will not need this for a SIP trunk, unless you want to maintain such calling habits to ease users transitioning from an old voice platform.

Scripts and Tools for Lync

- **Use Extensions**: This is another typical PBX feature with several variants. An extension is a short number (usually 3 or 4 digits) that allows users to contact any other company user directly. Extensions become less important on an SIP trunk or with the new Lync click-to-call feature, but it might still be useful to maintain this ability. We will take the most typical case of a user's extension that matches with the last four digits of the full number (DDI). Therefore, if your number is +12125551010, your extension will be 1010. Click on the **Edit extensions** button and configure our extension range with a rule suffix, main number prefix (that is, the "fixed" part of our number range, extension start (1001), and end (1099). In more advanced scenarios, multiple ranges can be defined.
- **Sip Trunk Connection**: By checking this box, no further number transformation will be performed before the call is sent out to the PSTN gateway. This is usually the case with SIP trunks that are capable of handling E.164 numbers with the **+** initial.

The following screenshot shows the parameters in the Lync Dialing Rule Optimizer:

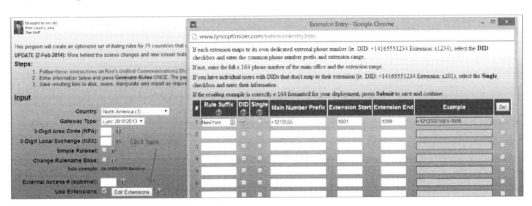

The following screenshot allows us to take a closer look at the form shown in the preceding screenshot:

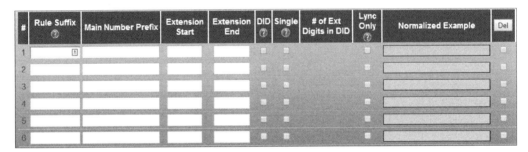

There's more...

All other parameters should be pretty self-explanatory, so change them as appropriate; now, click on **Generate rules**. The output will be a PowerShell script that you will need to download and copy on your Lync deployment to import the configuration. In order to run the script, open Lync PowerShell, browse to the folder where the script was saved, and execute it by typing its name.

Once done, you can enable your users for Enterprise Voice and assign the appropriate voice policy that the optimizer just created.

The author of the tool has provided detailed and comprehensive instructions on how to use the tool, with more examples and advanced usage at http://ucken.blogspot.co.uk/2012/01/complete-guide-to-lync-optimizer.html.

See also

- Although Dialing Rule Optimizer provides a great wizard-based solution for Lync voice configuration, some working knowledge of **Regular Expressions** (**RegEx**) is required. A good reference to review is http://msdn.microsoft.com/en-us/library/az24scfc(v=vs.110).aspx.
- For more detailed information about telephone numbering conventions and standards, including the E.164 format used by Lync, refer to *RFC 3966* at http://www.ietf.org/rfc/rfc3966.txt.

Switching between multiple Lync identities with a click – Profiles for Lync (P4L)

The Lync 2013 client has no built-in feature to save multiple profiles or switch between different identities; there are situations where users like Lync administrators might need the ability to quickly sign in with different accounts. **Profiles for Lync** (**P4L**) by Lync MVP Greig Sheridan (@greiginsydney) is a tool that will allow you to save multiple accounts and switch between them with one click, through a keyboard shortcut or through a command line.

Getting ready...

The tool is available at http://gallery.technet.microsoft.com/office/Profiles-For-Lync-2013-1b193329. This is the Lync 2013 version. A version for the Lync 2010 client is also available. Both the Lync 2013 and 2010 versions can be installed and coexist on the same machine.

Scripts and Tools for Lync

How to do it...

Using the tool is very simple. The following screenshot shows the main application window. Up to 40 identities can be saved. These are split across four tabs, each with its own shortcut key. The button highlighted in blue shows you which account is currently signed in; you can therefore switch identities in three different ways:

1. **INTERACTIVE**: Click on the number corresponding to the identity you want to use
2. **KEY SHORTCUT**: While the main utility windows is in the foreground, type the number corresponding to the identity you want to use (identities from 1 to 10 require no additional key)
3. **COMMAND LINE:** From a command prompt, browse to the folder where P4L is installed and type the following to activate profile no. 1:

   `"Profiles for Lync 2013.exe" /site=01`

Tracing made easier – Lync 2013 Centralized Logging Tool

One of the major Lync 2010 administrative shortcomings was the lack of a global logging facility. Server tracing required individually configuring, collecting, analyzing, and correlating separate logs from all involved machines. This turned into a considerably ineffective experience on larger deployments. Lync 2013 got a significant overhaul through the **Centralized Logging Service** (**CLS**). However, as it is only accessible through the command line (via `CLScontroller.exe` and other PowerShell cmdlets) and the many options around it, its usage is less than straightforward.

Centralized Logging Tool by Lync MVP James Cussen is a CLS graphical wrapper, which provides a far more intuitive yet cohesive logging experience with just a few mouse clicks.

Getting ready

The tool is available at http://www.mylynclab.com/2013/04/lync-2013-centralised-logging-tool.html. The author also provides comprehensive instructions and advanced examples; we recommend that you review the release notes in their entirety.

How to do it...

The script must be run from a Lync PowerShell. As it is not signed, you might first need to allow the execution of unsigned scripts. Proceed as follows:

1. Open PowerShell and type `Set-ExecutionPolicy Unrestricted`.
2. Run `\Lync2013CentralisedLoggingUI_v1.02.ps1`. Before the GUI shows up, Lync deployment will be scanned, as shown in the following screenshot:

```
Windows PowerShell
-------Getting current status of pools, please wait-------
Success Code - 0, Successful on 7 agents


Tracing Status:

DS-ECCPool2013.demo-suite.local (DS-ECCPool2013 v5.0.8308.0) (AlwaysOn=No)
     ds-ecc03.demo-suite.local (ds-ecc03 v5.0.8308.0) (Same as pool)
     ds-ecc02.demo-suite.local (ds-ecc02 v5.0.8308.0) (Same as pool)
DS-LED02.ds.ud-demo.com (DS-LED02 v5.0.8308.0) (AlwaysOn=No)
     DS-LED02.ds.ud-demo.com (DS-LED02 v5.0.8308.0) (Same as pool)
DS-LPC01.demo-suite.local (DS-LPC01 v5.0.8308.0) (AlwaysOn=No)
     DS-LPC01.demo-suite.local (DS-LPC01 v5.0.8308.0) (Same as pool)
DS-MED01.demo-suite.local (DS-MED01 v5.0.8308.0) (AlwaysOn=No)
     DS-MED01.demo-suite.local (DS-MED01 v5.0.8308.0) (Same as pool)
DSPool2013.demo-suite.local (DSPool2013 v5.0.8308.0) (AlwaysOn=No)
     ds-lfe03.demo-suite.local (ds-lfe03 v5.0.8308.0) (Same as pool)
     ds-lfe04.demo-suite.local (ds-lfe04 v5.0.8308.0) (Same as pool)
```

You might need to wait before the GUI shows up, while Lync Servers are queried for the status of the running logging scenarios (if any) and the state of each pool (multiple pools are supported). If a logging scenario is already enabled on the deployment, the tool will show it. The ability to execute specific actions will also be consistent with the detected state.

Scripts and Tools for Lync

Our test environment is composed of one Enterprise Edition Pool (two nodes) called `DSPool2013.demo-suite.local`, one Mediation Server, one Persistent Chat server, and finally one Edge Server, as shown in the following screenshot:

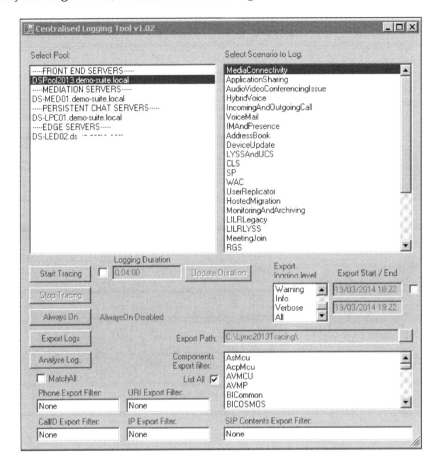

As you can see, in the top section of the preceding screenshot, you have an option to select which pool to log. Each core Lync role, irrespective of whether the pool is made up of single or multiple servers, is listed here consistently with the test scenario we described earlier.

In the top-right part, you need to select a scenario to log. There are many possible scenarios and the most correct one to choose depends on what type of component you need to analyze. For the purpose of our recipe, we assume that you want to log a media connectivity issue that occurs on the Front-End pool. Now, proceed with the following steps:

1. We will now want to log the `MediaConnectivity` scenario on the pool. From the **Select Pool** window, we pick the Lync Front-End pool, and from the **Scenario to Log** window, we select the scenario.

2. Now, click on **Start Tracing**. We will keep the tracing going for a couple of minutes and then we will click on **Stop Tracing**. In the real world, you will typically keep the trace going until the issue you want to troubleshoot is reproduced. The real-time status is displayed in the PowerShell, as shown in the next screenshot:

3. Now, click on **Export Logs**. This action will pull all individual logs from pool members, merge them in a single unified log, and export it to your designated file path, as shown in the following screenshot:

4. The output is a flat text file, which can be opened in a text editor but would be quite difficult to read. Like with Lync 2010, this file can be opened in the **Snooper tool**, which will make readability and event correlation a lot easier. You can do this directly from the logging tool by clicking on **Analyze Log**.

> The **Snooper tool** is now part of the Microsoft Lync 2013 debugging tools available at `http://www.microsoft.com/en-gb/download/details.aspx?id=35453`. For the **Analyze Log** button to work properly, Centralized Logging Tool expects it to be installed in the default location (`C:\Program Files\Microsoft Lync Server\Debugging Tools\Snooper.exe`). If this is not the case, then the Snooper path must be changed manually in the script. Refer to the author's blog on how to do this.

See also

Although the tool itself is quite self-explanatory, it requires you to already be familiar with *Lync Centralized Logging Service* concepts. For an overview of CLS and configuration scenarios, refer to the following links:

- Using CLS at
  `http://technet.microsoft.com/en-us/library/jj688101.aspx`
- Logging scenarios at
  `http://technet.microsoft.com/en us/library/jj688085.aspx`

Identifying recurrent issues – Lync Pilot Deployment Health Analysis

The **Lync Pilot Deployment Health Analysis Tool** is a lesser known Excel-based utility included in the *Microsoft Lync Rollout and Adoption Success Kit*. The tool retrieves and displays a summary of the key issues indicators, such as the most common failure code, in a simple way, serving as a precious complement (not a replacement!) to the most comprehensive information retrievable through Lync Monitoring Server and the related reports based on SQL Reporting Services. These are not only recommended but also required for the tool to work properly. Despite the pilot in its name, it is a valuable tool to provide a snapshot of the overall call quality even on aged production environments.

Getting ready...

The tool is a Microsoft Excel file called `Lync Pilot Deployment Health Analysis v1.1 (Lync 2013).xlsm`. It is available as part of the Microsoft RASK available at `http://www.microsoft.com/en-gb/download/details.aspx?id=37031`.

Ensure that the following prerequisites are met before using the tool:

- Microsoft Excel is installed on the machine from where you run the file
- The machine from where you run the file has connectivity to your Lync deployment
- The Lync Monitoring Server role must be deployed
- Call Detail Records and Quality of Experience policies must be configured to capture data

Retrieving report information requires some initial configuration. Start the file, navigate to the setup sheet, and configure the fields, as shown in the following screenshot:

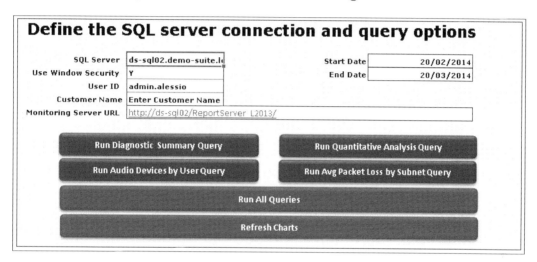

The following parameters need to be set:

- **SQL Server**: Use the FQDN\instance notation of the server where the Monitoring databases are placed, for example: `ds-sql02.demo-suite.local\LYNC`.
- **Use Windows Security**: Leave it at the default value.
- **User ID**: This refers to the user that will run the queries to monitor the databases; it therefore must be granted the minimum permission of `db_datareader` for the `LCSCdr` and `QoEMetrics` databases.

Scripts and Tools for Lync

- **Customer Name**: Enter your organization's name here. The name will be used on various generated charts.
- **Monitoring Server URL**: Input the base URL where your Lync monitoring reports are installed.

 Run Get-CsReportingConfiguration from a Lync PowerShell to retrieve the URL.

Make sure that you only include the base link to the reports server, as shown in the following screenshot:

How to do it...

Once a start and end date is configured and defined, you are ready to run the query of your choice (blue buttons) or red button. For our demo, please click on **Run all queries**.

 Use the **Run All Queries** feature with caution, especially if associated with huge date ranges and/or large deployments. Queries will need considerable time to run (Excel might become temporarily unresponsive), and this might affect the Lync SQL server's backend performance. It is sensible to run such extended queries after business hours, if required, and monitor SQL Server's performance impact for the future.

Among the several reports that you will want to review, two of these should be your earliest stops.

Chapter 5

The **Reliability Report** tab is a dashboard of the weekly session success rate for all core modalities (conferencing, peer-to-peer voice, and so on). It provides an immediate summarization of how the system is performing in the current week and how it did in the previous week, as shown in the following screenshot:

P2P Session	Number of Sessions	Session Success Rate		Weekly Change
		Current Week	Previous Week	
Application Sharing	1	**100.00%**	100.00% ⇨	0.00%
Audio	1233	**99.92%**	100.00% ⬇	0.08%
File Transfer	3	**100.00%**	100.00% ⇨	0.00%
IM	1283	**100.00%**	100.00% ⇨	0.00%
Video	2	**100.00%**	100.00% ⇨	0.00%

Reliability Levels			
Gold	Silver	Bronze	White
>=99.9%	>=99.5%	>=99.0%	< 99.0%

Related Monitoring Server Reports
Call Diagnostic Summary Report
Peer-to-Peer Activity Diagnostic Report
Conference Diagnostic Report
Top Failures Report
Failure Diagnostic Report

Scripts and Tools for Lync

The **Weekly Change** column indicates the change in the success rate from the previous week's results. This is a great indicator of the deployment health trend. Large shifts in values between two consecutive weeks should be promptly investigated. The **Diagnostic Comparison** tab should be used to find the frequency of specific error codes that affect the session's success rates. The overall performance quality is categorized in three color codes:

- Gold: This indicates that the success rate is greater than or equal to 99.9 percent
- Silver: This indicates that the success rate is greater than or equal to 99.5 percent
- Bronze: This indicates that the success rate is greater than or equal to 99.0 percent
- White: This indicates that the success rate is below 99 percent

There's more...

Your next stop should be the **Diagnostic Comparison** tab. This provides a more detailed view of the recurring failures that are affecting the platform.

> Information on specific diagnostic codes can be found in the `ms_diag.html` file, which is provided with the **Lync Server 2010 Resource Kit Tools**. The file is located in the `<Lync Resource Kit Installation Path>\Tracing\` directory. Alternatively, you can view them at http://msdn.microsoft.com/en-us/library/gg132446(v=office.12).aspx.

The **Diagnostic Comparison** tab is split into the following five main sections:

- **Goals**: This is where you set your acceptable failure rate for each modality (this defaults to 0.10 percent for unexpected failures). These values will define the purple goal line on each of the charts.
- **Classification Filtering**: This is used to break down to the exact percentage of each session result. You should emphasize the unexpected failures' results as they are health indicators of your Lync Server infrastructure. The data can be further filtered so that it only shows specific weeks and diagnostic codes.
- **Related Reports**: This links to the reports available for additional troubleshooting.
- **Charts**: These visually display the trending of the specific component's health throughout the date range specified in the **Setup** tab.
- **Date Drilldown**: This further filters specific error codes that might be causing spikes in the failure rates seen in the charts.

Chapter 5

The following screenshot shows all the parameters:

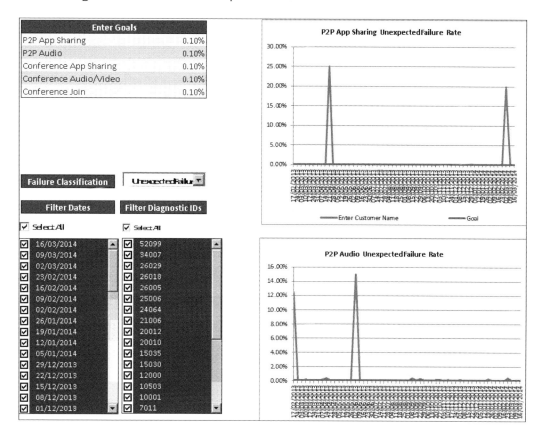

See also

- Microsoft provides more information about the tool usage on the `Lync Pilot Deployment Health Analysis Tool Instructions v1.1.docx` file included in the RASK.

Scripts and Tools for Lync

Managing phone numbers – Search-LineURI and Get-UnusedNumbers

If you have deployed or managed large Lync environments with Enterprise Voice, you should be well aware of how uninviting phone number management can be.

Assume you are running a large deployment with hundreds or thousands of number assignments and want to get a list of available or taken DDIs, or you need to assign a number to a new user and you need to look up the next available number in your range; or perhaps you attempt to assign a number, and you realize it's already taken, but you have no idea by whom. All of these supposedly normal administrative tasks will require either of the following:

- Advanced filtering on the Lync Control Panel
- PowerShell cmdlets or scripts
- An accurate and up-to-date documentation of number assignments

Search-LineURI (by Lasse Wedø) and **Get-UnusedNumbers** (by Ståle Hansen) are helpful tools, which will ease some number management aspects.

Getting ready...

The authors have started a joint project on `http://www.lyncnumbers.net` with the aim to collect tools and information about Lync number management. You will find their updated scripts available as individual downloads, specifically **Search-LineURI**: `http://gallery.technet.microsoft.com/Search-for-that-LineURI-814ac281/` and **Get-UnusedNumbers**: `http://gallery.technet.microsoft.com/Find-next-available-number-58391c72`.

How to do it...

Search-LineURI is a PowerShell script that will search and list/display all numbers assigned on a Lync environment. It includes numbers assigned to users as well as special resources such as Response Groups, common area phones, and so on; all resources that can have a number assigned are covered. It comes with several switches and options and we recommend that you review these in their entirety by opening the PS1 script in a text editor and reading through the description. We will look at a practical usage example along with the most useful options.

Assume that our Lync 2013 Enterprise Voice deployment uses the following numbering range:

- The first number is 212-555-1001 (in E.164, this is +12125551001)
- The last number is 212-555-1999 (in E.164, this is +12125551999)

Chapter 5

The script must be run from Lync PowerShell. As it is not signed, you might first need to allow the execution of unsigned scripts. Proceed as follows:

1. Open PowerShell and type `Set-ExecutionPolicy Unrestricted`.
2. Run the script as follows: `.\Search-LineURI.ps1 -search 5551 -ListAsGridView`.

The `-search 5551` (actually, it is *5551*) option tells the script to match all numbers that contain the 5551 string in the URI. It should be noted that a minimum of 2 digits must be provided in the search option.

The `-ListAsGridView` command instructs the tool to display a GUI-based list, as shown in the following screenshot:

LineURI	PrivateLine	DisplayName	Name	SipAddress	Identity
tel:+12125551001		Han Solo	Han Solo	sip:Han.Solo@giombini.com	CN=Han Solo,OU=TestReso
tel:+12125551002		Leia Organa	Leia Organa	sip:Leia.Organa@giombini.com	CN=Leia Organa,OU=TestR
tel:+12125551003		Luke Skywalker	Luke Skywalker	sip:luke.skywalker@giombini.com	CN=Luke Skywalker,OU=Te
tel:+12125551004		Obi-Wan Kenobi	Obi-Wan Kenobi	sip:obi-wan.kenobi@giombini.com	CN=Obi-Wan Kenobi,OU=T
tel:+12125551999			RG1		service:ApplicationServer:lyr

The output will list all assigned numbers (for users, it will display both the main line URI and the private line, if applicable) along with the SIP address and identity path. Note that the last entry (name: RG1) actually belongs to a Response Group.

Other typical usages are described in the following table:

Command	What does the command do?
`.\Search-LineURI.ps1 -search 5551`	This finds all assigned numbers that contain 5551 in the URI and displays them within the PowerShell session.
`.\Search-LineURI.ps1 -search 5551 -savetoxml`	This finds all assigned numbers that contain 5551 in the URI and saves the output at `C:\Lync Documentation\Search-LineURI-Result.xml`. The path and the filename can be customized with the help of additional options.

Using the Get-UnusedNumbers script

The **Get-UnusedNumbers** script is a PowerShell script that will search through all users, special services, and devices and list/display unassigned numbers.

Scripts and Tools for Lync

For proper operation, the script needs to know what number ranges are being used on the Lync platform; the following are the two ways to achieve this:

- Leverage the **Unassigned Number** voice feature in Lync. You will need to configure this before running the script.
- Specify a custom range within the script

Let's make a practical example for case 1, assuming our Lync deployment has a SIP trunk with the following numbering range:

- First number: 212-555-1001 (in E.164: +12125551001)
- Last number: 212-555-1019 (in E.164: +12125551019)

We first need to configure an announcement for unassigned numbers. A full explanation about all options is available at `http://technet.microsoft.com/en-us/library/gg425944.aspx`. For our example, we will configure a simple text-to-speech announcement through the following PowerShell command (note, announcements cannot be configured through the Control Panel):

```
New-CsAnnouncement -Identity applicationserver:lyncsefe01.giombini.com -Name DefaultAnnouncement -TextToSpeechPrompt "This number is not assigned" -Language en-US
```

The result is shown in the following screenshot:

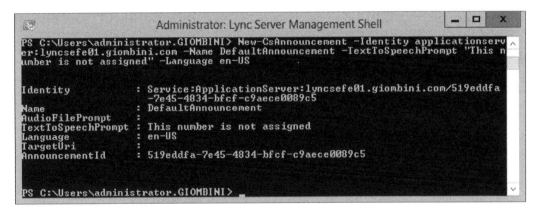

We are now able to configure an unassigned number range. This can be done through PowerShell or the Control Panel. We will use the latter and configure as per the following illustration (when done, click on **OK** then the **commit** button):

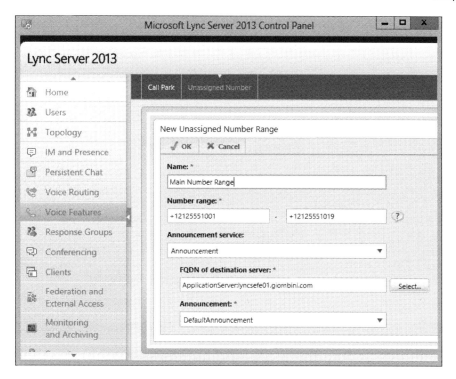

Now, we need to change something in the script before being able to run it interactively, and leveraging the unassigned number feature. Please open the script with a text editor (Notepad or the like), and browse towards the end until you find the following line:

```
$FirstNumber=Get-Unused -RangeStart +47232323001111 -RangeEnd
+47232323001190 -Name "Custom Range Norway" #-ListAll $true
```

Change the preceding lines to the following:

```
$FirstNumber=Get-Unused -Name "Unassigned Number range" -ListAll $true
```

Such change will instruct the script to not use a custom phone range (therefore it will use the range defined in the unassigned number announcement feature), and will list all free numbers, not just the first one.

We are now ready to run the script by opening a PowerShell, browsing to the folder where the script was saved and typing the following:

```
.\Get-UnusedNumbers.ps1
```

Scripts and Tools for Lync

The first useful information we retrieve is our range is made of 19 numbers, 14 of which are not assigned. Then, type L, to list all available numbers, and then press Enter. This will return all unused numbers in our deployment, as shown in the following screenshot:

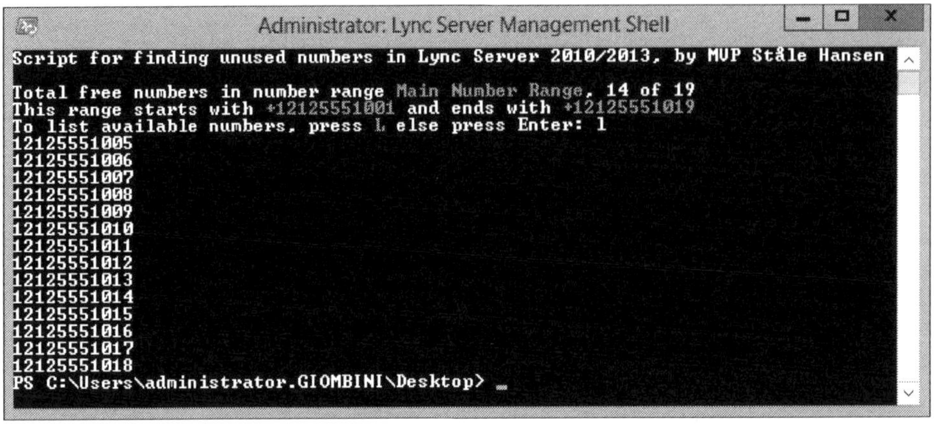

Now, let's make another example for case 2. This time, we will not be using the unassigned number feature; we will instead specify a custom parameter in the script. Please go back to the following line in the original script:

```
$FirstNumber=Get-Unused -RangeStart +47232323001111 -RangeEnd +47232323001190 -Name "Custom Range Norway" #-ListAll $true
```

And change the preceding line as follows:

```
$FirstNumber=Get-Unused -RangeStart +12125551001 -RangeEnd +12125551009 -Name "My Custom Range" -ListAll $true
```

We are now ready to run the script by opening a PowerShell, browsing to the folder where the script was saved and typing:

`.\Get-UnusedNumbers.ps1`

The first useful information we retrieve is our custom range is made of 9 numbers, 5 of which are not assigned. Then, type L, to list all available numbers, and then press Enter. This will return all unused numbers in our deployment (this time, in our custom range defined within the script):

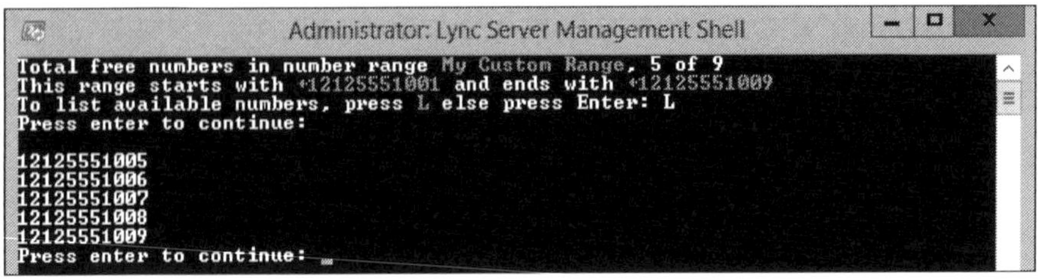

The currently reviewed version is v.4 and does not feature the ability to export the output to a table or a file. This is expected in a future version.

Another great feature is the script is a function, and may be then reused in other scripts. An example could be reusing the function in a script that enable users for enterprise voice, retrieve the first available number in the range, then programmatically assign it to users as they get automatically enabled. This is also the reason why we needed to change some parameters directly within the script before being able to run it interactively. The author is planning to add support for parameters in a future version.

Managing Call Pickup Groups – Lync2013CallPickupManager 1.01

Call Pickup has been, among many others, a long established voice feature of almost any PBX. A **group call pickup** is a logical aggregation of voice users who have the ability to pick a call directed to any group member by dialing a special access code assigned to the group.

For example, assume that John and Mary are Lync Voice users and are members of a pickup group called SalesTeam, and the access code is #101. Mary receives a call on Lync, and her desktop phone is ringing, but she is not at her desk. John can dial #101 on his Lync endpoint and answer the call on her behalf.

Group call pickup was introduced in the February 2013 Cumulative Update. Full details on how to configure it are available at Microsoft TechNet (http://technet.microsoft.com/en-us/library/jj945640.aspx). However, that's another example of less-than-optimal integration, as leveraging this functionality will require the command line, including the **SEFAutil** tool that is included in the Microsoft Lync 2013 Resource Kit Tools.

Lync2013CallPickupManager is a GUI wrapper, which will make group pickup management significantly easier and more cohesive.

Getting ready...

The tool can be downloaded from the author's blog at http://www.mylynclab.com/2013/10/lync-2013-call-pickup-group-manager.html. As mentioned before, it leverages the SEFAutil component for proper functionality. These are the steps required to get it up and running:

1. Ensure that your Lync 2013 infrastructure has been updated with the February 2013 Cumulative Update (CU1) or a later version. The latest CU is highly recommended.

2. On the server where you plan to run the tool, install **Microsoft Unified Communications Managed API (UCMA) 3.0 Core SDK** or later. Version 4.0 can be downloaded from http://www.microsoft.com/en-gb/download/details.aspx?id=34992. Note that servers where Lync roles are installed already have this component installed.

Scripts and Tools for Lync

3. On the same server, download and install **Lync 2013 Resource Kit Tools**: http://www.microsoft.com/en-au/download/details.aspx?id=36821.

4. Configure a trusted application pool for SEFAutil. Run the following:

   - `New-CsTrustedApplicationPool -id <Pool FQDN> -Registrar <Pool FQDN> -site Site:<Site>`. Replace `Pool FQN` and `Site` with your actual values
   - `New-CsTrustedApplication -ApplicationId sefautil -TrustedApplicationPoolFqdn <Pool FQDN> -Port 7489`
   - `Enable-CsTopology`

How to do it...

You are now ready to use the tool. Let's run it and configure a sample scenario. When the application starts up, it will check whether the prerequisites are satisfied, and it will scan the Lync infrastructure, as shown in the following screenshot:

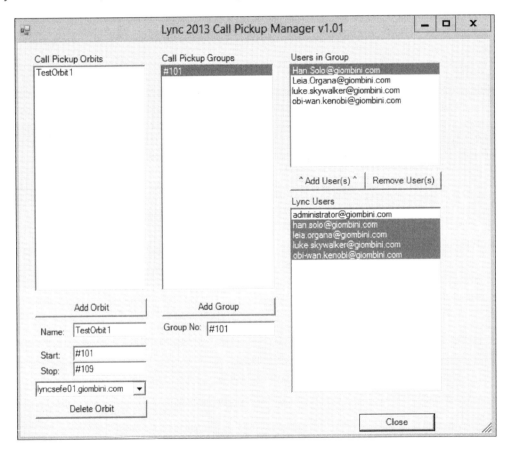

The first element that we need to configure is a **Call Parking Orbit**. This is essentially a range of special-purpose codes, each of which is associated with a pickup group, and corresponds with the number that member users should dial to pickup the call. In our example, we have assumed that we need nine pickup groups, so we have configured the following:

- A name for the orbit (**TestOrbit1**)
- A start code (**#101**)
- A stop code (**#109**)
- The Lync Pool where we configure the **Call Pickup Group**

We then clicked on the **Add Orbit** button. Note that we have used **#** as a special purpose character to avoid possible overlapping with extensions.

We are now ready to add members to our first group. A call pickup group is actually named after its associated pickup code; in our case, it's **#101**. Once you add the group using the **Add Group** button, you will be able to select the available users and join them to the pickup group. You will notice that the tool will execute the SEFAutil and perform the requested operation. Note that this might require some time as the join command needs to be reiterated for each user.

> Importantly, a user can be a member of only one call pickup group at a time.

6
Designing a Lync Solution – The Overlooked Aspects

In this chapter, we will cover the following recipes:

- Meeting your users' expectations
- User training
- Gathering users' requirements
- Weighing up around Lync virtualization
- Network readiness – introduction
- Defining personas for the network
- Defining sites for the network
- Network readiness – review and analyze results

Introduction

Microsoft Lync Server 2013 is the sixth generation of an increasingly popular universal communications and collaboration platform. It provides real-time communication and collaboration features such as presence, instant messaging, native Voice over IP audio and video calling, conferencing, application and desktop sharing; Lync also serves as a full PBX and telephony replacement through integration with PSTN and SIP networks (which Microsoft refers to as Enterprise Voice).

A great deal of documentation and tools exist on how to implement key elements of a successful Lync solution (including this very book). At face value, Lync installation seems like an easy, wizard-based, repetitive process. What many seem to overlook is the impact of adding real-time communications to an existing network infrastructure, an issue that we have seen occurring far too often.

The reality is that Lync is a complex ecosystem with several dependencies, some of which are not strictly technical.

Deploying certain features, such as voice and video (full HD is supported on Lync Server 2013), clearly has a considerable impact on any existing IT infrastructure, starting from the lower OSI layers. Designing a stable and capable Lync solution is not a simple task, and there is no way you can expect to deliver a successful implementation without accurate knowledge and assessment of existing IT infrastructure, planning, and design, and accurate plans for maintenance and upgrades. Nevertheless, even after following a technically impeccable deployment, your project might fail if you overlook some of the key elements, which we aim to describe here. This chapter will give you some insights on fundamental but often overlooked elements of the design, implementation, and ultimately, success of a Lync solution through a mix of technical and non-technical recipes.

Meeting your users' expectations

A typical but potentially major mistake in the design of a Lync solution can be summarized in two words: **human factor**.

It is typical among technical decision makers, especially those from hi-tech firms, to put excessive bias on technological aspects while overlooking or forgetting the end users. Specifically, we refer to the following points:

- **Expectations**: The ability for your solution to meet their demands/requirements
- **Acceptance**: Involving users in the decision-making process, and allowing for their opinions to influence the final solution
- **Experience**: Envisioning a simple and effective solution that can suit people with diverse skills and culture
- **Knowledge**: Making people aware of your new UC solution, and training them to maximize their productivity and effective communication

Technology definitely matters, and if your infrastructure fails to deliver a consistently good Quality of Experience, the users will abandon it, which will render your investment void.

The way to meet users' expectations and deliver on requirements starts with awareness. Making users aware of what is coming will somehow make them feel involved in the process and will play a fundamental role in the success of the project.

We can list the challenges you will face in the following two broad categories:

- **User readiness**: Have you understood and addressed the user requirements in your design? Note that user requirements are not necessarily a synonym of company requirements.
- **User adoption**: Have you ensured that a consistent part of your project will be devoted to user awareness and training? Lync should not be just another icon on their desktop.

Unified Communications are, in many cases, a significant shift from established communication patterns and habits, which will definitely affect productivity (for better or worse, it is partly up to you). Although instant messaging and other successful technologies have long gained traction and popularity among users, a significant number of companies still rely heavily on traditional forms of communication such as PBX-based telephones and e-mail. The transition from tradition to innovation (or convergence of diverse communication technologies into a more homogeneous ecosystem, which is a UC pillar) is something you will need to encourage and support.

How to do it...

When planning for requirements, treat all business units within the company as different realities, each with their unique view, and do your best to correlate the different requirements into an organic solution. We suggest that you compile a list of requirements for each unit, and encourage your customers' key decision makers to do the same. Their continued involvement in the project is fundamental. You may then correlate the lists in a global repository and start thinking about technical challenges.

A typical example of a biased approach is only listening to IT teams and technically educated users. An amazing number of enterprises do not conduct proper end-user surveys, leaving it to IT teams or key decision makers (which, not surprisingly, often do correspond) to lead the process, sometimes with a disappointing outcome for the real utilizers.

When you conduct interviews with end-user representatives, and as the requirements list starts building up, ensure that you consider and address all of the following aspects:

- **Environmental (globally)**: Think about global companies with worldwide presence and their different culture, technical readiness, habits, and legislation. For example, some places might culturally prefer different forms of communications over others, and you will need to take this into account when estimating the capacity calculations for network bandwidth, server placement, user policies, and so on. We stress on the importance of not forgetting to consider all departments, and ensuring that you get attendance from key stakeholders for each. Not all of them will have the same requirements. For example, some might put greater emphasis on video communication (and perhaps demand advanced devices such as Lync Room systems) and indicate it as a key business requirement. This would result in fundamental information to help shape the network capacity to accommodate such needs.

- **Productivity and comfort for users**: As we said before, the new communication workloads that are typical of a Lync solution may require some time for proper adoption. Ensure that you address transition properly. Advanced Lync users usually put greater emphasis on software-based endpoints; they will mostly be happy and fully productive with their Lync client and a good (read certified) pair of headsets; this might not initially apply to those who are used to traditional desktop phones. There is a plethora of Lync-compatible devices from several partners that are easy and intuitive to use, yet provide old-school digit-based dialing, along with Lync-native advanced features. Products range from aggressively priced entry-level models to high-end devices. On the telephony note, also remember to address power-users such as receptionists and attendants (a frequently overlooked category), including specific training. Being usually the first and most important ingress point for business calls, it is essential that you meet their habits. Microsoft provides a Lync Attendant Console, which is a good fit for simpler scenarios and call flows. The console dates back to Lync 2010 (but it is compatible with 2013). For more advanced scenarios, there are excellent third-party solutions (which may suggest why Microsoft relied on partners' solutions and never updated its 2010 console).

- **Requirements can pose challenges to technical and budget readiness**: A typical occurrence is developing specific workloads that depend on high-quality networks in less developed countries (voice and video), or remote sites where degraded network performance to a far central site is a physiologically inevitable occurrence. Make sure you have a very clear understanding of users' breakout, and compile a list of sites, each with the user count, total network bandwidth, available bandwidth for RTC, requirements for Call Admission Control, and **Quality of Service** (**QoS**). Such information will be fundamental in determining the network readiness for your deployment (we will cover this aspect in *Chapter 11, Controlling Your Network – A Quick Drill into QoS and CAC*).

- Legislation will play an important role in some of your design decisions, as you may be required to deploy special compliance features (such as IM and conferencing archiving, which Lync provides out of the box) or possibly a global call record solution (which includes audio and video calls, whether PSTN or Lync-native). Lync 2013 can provide limited call recording capabilities, which were clearly not designed with legal compliance in mind (that is, recording is manually started by a participant in the conference, and this cannot be managed centrally. Furthermore, non-conference PSTN calls cannot be recorded unless through a workaround). A fully manageable call recording for compliance requirements will require third-party solutions. Voice over IP and PSTN regulations are also likely to affect your global design. Some countries may not allow IP-based voice, so your only option here may be deploying one or more PSTN gateways through a local trunk; some others place specific restrictions like only allowing national carriers, or prohibiting **Voice over IP** (**VoIP**) outbound traffic to be routed outside the national boundaries unless through a national carrier.

- **BYOD (Bring Your Own Device)**: This is a ramping-up trend, which, in many cases, does not play well with organizations subject to rigid security requirements. Lync 2013 mobility now brings full audio and video capabilities to mobile devices, and this is getting a lot of attention from businesses that see it as an attractive productivity booster or cheap resiliency solution for branch offices without a redundant network link or a local Lync Survivable Branch Appliance. Think about a site with a failed WAN link to the central Lync site. Mobile devices would still be able to connect through a local Wi-Fi gateway (or even through the 3G/4G data network) via the Lync Edge server, and be able to make and receive calls through the corporate phone numbers.

There's more...

In this recipe, we have provided several, but not all, examples of what type of user-centric challenges you would face in a Lync project. The two key take-away concepts are:

- Allow technological decision to be driven by user expectations and not exclusively the other way round
- Adopt a user-centric approach as a fundamental stepping stone as this is essential for the type of business transformation and project success many companies aim for when it comes to deploying a Unified Communications solution

See also

- For more information about Voice over IP legislation around the world, refer to `http://www.absoluteuc.org/can-use-lync-client-ip-telephony-voip-regulations-around-world`.
- Use of Lync-certified devices, including headsets, plays an important role in the quality of experience. For a catalog of Lync devices, refer to `http://catalog.lync.com/en-us/hardware/index.aspx`.

User training

We briefly introduced user-training concepts in the previous recipe, but now, we would like to expand on such a fundamental milestone for every Lync project's success.

In the eyes of non-technical users, the first impression when you run the Lync client is that of just another nice-looking instant messaging software. Although several features are intuitive (and relatively simple to learn and manage), other advanced functionalities sit somewhat in the background and are not immediately evident to the uneducated user. A thorough understanding of these would require brainstorming and adequate training.

The difference between trained and untrained users can be huge, and will be a major differentiating factor in the success of a Lync implementation.

How to do it...

Now, it is time to break down the main aspects you will need to cover when planning for user training, along with the benefits for each:

- **Manage the change**: This is a typical challenge when rolling out a new technology, but is often overlooked. Not all people react to changes in a positive way. Highlighting Lync's features and improvements over traditional communication will be ground for improved acceptance and to establish habitual usage.

- **Reduce escalation**: Incomplete knowledge will inevitably result in repeated and redundant calls to your user support for requests that should be managed elsewhere. You do not want your employees to flood support with requests such as "how do I make a call", "how do I add a contact", or "how do I use Lync from my smartphone". Such occurrences are a clear indicator that you have not addressed user knowledge properly, and they are a threat for your support engineers' ability to manage "real" issues effectively.

- **Make your users advanced utilizers**: There is no such thing as hidden features in Lync, but some are less evident, or are part of the "unified communications" revolution that users may be less accustomed to. Be sure to include this in your training plan; it will make a huge difference.

- **Presence**: This is a fundamental pillar of UC. Ensure that your users understand it and manage it properly. Explain the impact of the presence status, how these are managed (for example, presence changes automatically based on calendar entries, or when we are in a call, and so on), and what is the impact of each presence status for users' ability to communicate. For example, a user whose state is "in a call" would most likely be able to accept IM, whereas a user that does not want to receive communication should set the "do not disturb" presence status. It is also important to make users aware of device impact (*no IM – voice only* indicates that the user is signed in on a Lync phone, but this is frequently ignored).

- **Privacy relationships**: This is a frequently ignored Lync feature but can have great relevance as a basic ethical firewall. Privacy relationships control what part of presence and contact card information are exposed to other users. Each contact in the list can have five different privacy relationships with you, each of which provides different access levels.

- **Use the right communication feature**: As part of a broader concept of optimal infrastructure usage and netiquette, educate your users on how to use the right feature to communicate. For example, IM before calling, respect presence status like "busy", use a Lync call in-lieu of a PSTN call (no toll charge), and so on.

- **Advanced calling features**: Making and receiving a basic call is intuitive, but how about voice features such as call park, transferring a call, setting up call forwarding, dial-out a user in a conference, and team calling? These have a great impact on user and team productivity. You must ensure that you cover specific training for this.

- **Improve productivity and ROI**: Put a lot of emphasis on how Lync improves the communication for people on the move, and how this translates to considerable cost savings due to saved traveling and call charges.
- **We are not all equal**: Do not forget to address specific training requirements for special user categories such as receptionists and switchboard operators. Special users are also executive assistants and those who frequently manage communications on behalf of others.
- **Last but not least, IT staff training**: Unless you outsource your customer support, your IT teams will require specific training to be able to handle end-user and infrastructural issues.

Depending on your company size, you might face several challenges while planning the training, such as what content should you deliver? How would you make content available? Is classroom-based training a feasible option, or can you just rely on remote/self-paced training? Different backgrounds, different usage, and dispersed teams will need to be properly handled. You should of course consider Lync as a great training solution itself.

There's more...

Although there is no one-size-fits-all solution, we will start from a common ground of providing good resources for training material:

- Lync instructor-led training (http://office.microsoft.com/en-us/lync/lync-instructor-led-training-HA102465959.aspx): These are free sessions with an instructor on a wide range of topics from basic usage to advanced users, including voice and conferencing.
- The Microsoft Lync 2013 rollout and adoption success kit (available at http://technet.microsoft.com/en-us/lync/jj879331) is a one-stop resource for excellent training material, including:
    - Getting-started guides
    - Quick reference cards

The full list of training resources is found in the `Lync_User_Education_Training.xlsx` Excel spreadsheet in the RASK.

Gathering the users' requirements

Requirements gathering processes should be subject to a well-structured process. Typically, it all starts with commercial engagements, which, most likely, you will not be involved with, right from the start. However, Unified (or Universal) Communications are a vastly complex world that require considerable skills, and you should expect the sales overlays to involve technical resources in the discussion quite early. This is where you will step in and expect to become the enabler for several key aspects in the project's life cycle.

Getting ready

Collecting requirements and translating/correlating these into an infrastructural design and project plan will require considerable time and effort. It is best to stay organized and rely on resources and documents, which will help keep the information in good order.

If you are a Microsoft Partner, an invaluable resource to help organize your work is the Microsoft Lync Deployment Planning Services (available at `http://planningservices.partners.extranet.microsoft.com/en/edps/pages/default.aspx`). Here, you will find excellent material to manage your engagement, from design and Bill of Material questionnaires to planning the solution.

How to do it...

You will need to get a good understanding of your customer's requirements by organizing meetings and workshops with key stakeholders. Remember, do not just rely on technical people! Ensure you get attendance from every relevant business unit. Each will have unique requirements, which you should try to fit in your solution. Listen to everyone! A successful Lync project will ultimately depend on user acceptance and adoption (we will further expand this concept in a following recipe).

Among the key aspects, you will need to cover the following in your sessions:

- **Understand customer project drivers**: Why do they want to implement Lync? What benefits they aim to achieve?
- Are they currently undergoing an evaluation process with other competitors' solutions? Alternatively, are they considering Lync as a potential replacement for an existing UC solution that failed to meet their demands?
- **Understand their business requirements**: How do they currently communicate? Do they think their communication culture is open and efficient? How would they benefit from new ways of collaborating?
- **Understand customer issues**: Highlight how Lync might be an effective and simple solution to consolidate siloed and disjointed communications platforms, including traditional PSTN telephony, into a simpler and integrated ecosystem, ultimately enhancing effectiveness, administration, and return of investment.
- Introduce the customer to the features and functionality of Lync Server 2013 and map them to their issues and requirements.
- Understand and discuss technical requirements, assessing if a customer's underlying infrastructure is ready for deployment or may require upgrades.

- Help the organization to develop a deployment strategy for their organization, in the form of a solution concept, and envisage a staged roadmap to add more features and support more users in future phases. Although many sources state "pilot" phases as risky quicksand, if well conducted, they are an invaluable resource to prove the great benefits of Microsoft Lync "hands on".

- Provide the opportunity to leverage a Lync expert for their deployment planning activities and questions. Many technical customers, although most likely not entirely UC-knowledgeable, are aware of potential impacts associated with deploying real-time communications, and they will want to discuss it. This is especially key when Lync is not standalone but is integrated with PBX/PSTN, third-party products, alternative video-conferencing solution, and so on.

Weighing up around Lync virtualization

The long-established trend to virtualize resources might suggest that applying virtualization, whether based on VMware or Microsoft Hyper-V, to a Lync infrastructure could be the likely, if not obvious, choice.

This is especially true when virtualization has already undergone successful adoption by your customer, and their IT teams are pushing for it. This also applies to UC partners who sell cloud-based Lync solutions, which are either single or multi-tenanted.

Virtualization is another among the many overlooked aspects that has led many Lync projects to the risk of failure. It should be noted that early documentation about Lync virtualization requirements was somewhat vague. However, the release of the *Lync Server 2013 Virtualization* white paper filled this gap.

The main issues that arise from virtualization relate to the system's performance. Virtualization plays an important role in cost optimization by leveraging hardware resource-sharing techniques among virtual servers, so that a physical host may run more virtual servers than what the nominal hardware specifications might suggest.

Lync Server, due to its inherent nature of real-time communications platform, however, is more sensitive to computational resource shortage (even temporary) than most other applications and requires special treatment. A temporary performance issue on an e-mail server application would most likely be unnoticeable to users; the same situation on Lync would have a potentially disruptive effect on users' experience, including calls unexpectedly being terminated, poor audio quality (audio dropouts similar to poor mobile coverage), and so on.

Lync virtualization, therefore, demands for very specific planning and configuration, which we aim to describe in the next section.

Designing a Lync Solution – The Overlooked Aspects

How to do it...

We may be pointing out the obvious, but we assume you will be using a certified Hypervisor for your deployment. The list of supported and/or validated virtualizing infrastructures is included in Microsoft SVVP supportability pages available at http://www.windowsservercatalog.com/svvp.aspx, and includes, among others, the following:

- Microsoft Hyper-V (but not all versions)
- VMware

> Each supported hypervisor might have environment-specific requirements, which are out of scope here, but we recommend that you review them at the relevant vendor's support pages.

The first item you must address is **planning for server capacity**; you will need to ensure that you assign the correct hardware resources to your virtual machines based on Microsoft's specifications, remembering an additional overhead of around 5% to 10% above what an equivalent physical server would require.

Microsoft's recommended requirements for Lync Servers are the following.

For Lync Front End, SQL Server (Lync backend), and Persistent Chat servers:

HW component	HW requirement
CPU	64-bit dual processor, quad-core, 2.6 GHz or higher
Memory	32 GB
Disk	One of the following: - 10K RPM hard disk drive (HDD) - High-performance solid state drive (SSD) with performance equal to or better than 10K RPM HDD - 2x RAID 10 (striped and mirrored) 15K RPM disk set
Network	Two interfaces required, either one 2-port 1 Gbps NIC or two 1-port 1 Gbps NICs.

For Lync Edge, Director, and standalone Mediation Servers

HW component	HW requirement
CPU	64-bit dual processor, quad-core, 2.6 GHz or higher
Memory	16 GB
Disk	One of the following: ▶ 10K RPM hard disk drive (HDD) ▶ High-performance solid state drive (SSD) with performance equal to or better than 10K RPM HDD ▶ 2x RAID 10 (striped and mirrored) 15K RPM disk set
Network	Two interfaces required, either one 2-port 1 Gbps NIC or two 1-port 1 Gbps NICs.

Refer to the TechNet article at http://technet.microsoft.com/en-us/library/gg398835.aspx for more details on hardware requirements.

Smaller environments might call for some slight adjustments on the hardware specifications, as some sources may suggest, but we recommend that you stick with Microsoft's recommended specifications.

You will also need to apply the following other specific hardware-related configuration:

- Hyperthreading should be disabled.
- Do not oversubscribe the CPU (1:1 ratio is required).
- Ensure that the host servers support **nested page tables** (**NPT**) and **extended page tables** (**EPT**).
- Disable **non-uniform memory access** (**NUMA**) spanning on the hypervisor.
- Do not configure dynamic memory or memory over commitment on host servers.
- Use **Virtual Machine Queue** (**VMQ**) to optimize synthetic NIC performance.
- Use physical NIC segregation (do not share the same physical NIC among several virtual servers).
- **Single-root I/O virtualization** (**SR-IOV**) is recommended. The specific configuration you should use depends on the host chipset and network adapter/driver.
- Use fixed-size or pass-through disks rather than dynamic disks. Dynamically-expanded disks are not supported on Lync Server.

Secondly, you need to plan for the virtual machines' placement. On a Lync solution that employs multiple-node pools for resiliency (Enterprise Front End pool, Edge pool, Mediation pool, and so on), it is essential that some servers in the pool remain functional for service continuity. When planning your Lync Server's placement, ensure that your virtualizing infrastructure can honor the Lync resiliency model so that you can split virtual servers consistently. Some examples of poor planning are as follows:

- Your Lync Enterprise pool is composed of two virtual servers, and you place them on the same virtualizing host (a failure in a host component will cause the entire Lync pool to fail).
- Elements in your virtualizing infrastructure are single points of failure. Examples are a single physical host; a single critical component, such as physical network cards; non-redundant storage; non-redundant Power Switching Units; and so on.

Action: split Lync virtual servers across multiple physical hosts; make sure there is no single point of failure in each component, and configure anti-affinity rules so that Lync Servers with identical roles are not hosted on the same physical server.

Third, you should consider resiliency. Lync Server 2013 and its related dependencies such as SQL Server come with its own set of native software-level resiliency techniques, such as:

- Data replication across multiple servers in the same pool
- Pool pairing
- For a SQL server dedicated backend, mirroring, or failover clustering

Virtualization software has also long employed availability features such as Live Migration (Microsoft Hyper-V) or vMotion (VMware), and virtual machines' snapshots. These techniques allow a live virtual server to be moved from one physical host to another without disruption of service availability. Snapshots can be used to restore a virtual machine to a given state. Although it might be tempting to use these with Lync Server, none of them are supported because there is currently no technical solution to move an active media session without appreciable impact to users (maybe in the future). At present, Lync VMs must be shut down before they can be moved. In addition, restoring a virtual snapshot of a Lync Server will have completely unpredictable and, most likely, adverse results.

Action: ensure no vMotion or Live migration is configured on Lync VMs. Do not use virtual snapshots as a disaster recovery method for Lync Server.

Fourth, you should plan for **resource monitoring**. Because of the additional layer of hardware abstraction adopted through virtualization, resource monitoring is more complex. On a deployment based on physical servers, you would typically monitor resources such as performance counters and disk I/O; on a virtualized environment, additional monitoring will be required on the physical hosts so that, in case of performance issue, the root cause can be narrowed down to the physical or virtual layer.

There's more...

The decision about whether to take the virtualization path or not is a fundamental design decision that should be made in the very early project phases.

You may expect the team that is managing the virtualizing infrastructure to object to Microsoft's prescriptive guidance and requirements to virtualize Lync Server, and insist on leveraging features such as hardware resource sharing/oversubscription, dynamic virtual machine migration, and so on.

Furthermore, the inability to use advanced features may impair virtualization's operational efficiency and render the virtualization option less attractive to a business, due to the requirement of dedicating host resources to Lync virtual servers. It will be fundamental here to consider a cost/benefit analysis.

See also

Planning a Lync Server 2013 Deployment on Virtual Servers is a fundamental read. It provides guidance to deploy Lync Server 2013 on virtual servers. It includes recommendations for the configuration of host servers and guest servers, key health indicators to watch during testing and deployment, and observations from Microsoft performance testing of Lync Server 2013 in a virtual environment. It is available at `http://www.microsoft.com/en-us/download/details.aspx?id=41936`.

Network readiness – introduction

We briefly introduced the importance of assessing the underlying network to ensure it is capable of handling real-time traffic for UC communications. In this recipe, we will present a practical example of network readiness assessment.

Getting ready

To prepare for the assessment, we need the following two main items:

- The Lync 2010 and 2013 Bandwidth Calculator (available at `http://www.microsoft.com/en-us/download/details.aspx?id=19011`). This tool is a Microsoft Excel spreadsheet; it was purposely built to calculate bandwidth requirements based on estimated usage, scenarios, and network parameters.
- A network assessment with the following information:
    - A list of sites where Lync endpoints will be located.
    - A count of Lync users for each site.
    - Nominal WAN bandwidth for each site (that is, maximum theoretical bandwidth).

- Available bandwidth for RTC (that is, the amount of bandwidth we can reserve, in conjunction with QoS, for some Lync traffic).
- Bandwidth of the central site, that is, a site where Lync Servers are located. In our example, this will be the company's datacenter.
- An estimation of traffic usage by modality. This will be the baseline to build our personas. We will get back to this shortly.

How to do it...

For our test scenario, we will be using a fictitious company with the following parameters:

Site	Users	Maximum Bandwidth (Mb)	Available RTC Bandwidth (Mb)	Description
Datacenter	0	1024	200	
New York	4000	100	20	
Chicago	1000	100	20	
Dallas	500	20	4	
Denver	500	50	10	
Miami	100	10	2	
Reno	50	4	1	
Los Angeles	30	4	1	

You might notice that the *Available RTC Bandwidth* column value is 20 percent of the maximum bandwidth for each site.

The other parameters we will use in our demonstration are as follows:

- **Quality of Service** (**QoS**) is described in *Chapter 11, Controlling Your Network – A Quick Drill into QoS and CAC*
- **Call Admission Control** (CAC) is also described in *Chapter 11, Controlling Your Network – A Quick Drill into QoS and CAC*
- There is a single central PSTN gateway in the data center site

Chapter 6

It is now time to start the Bandwidth Calculator tool and set up a few preliminary parameters. You will notice that the tool is pre-populated with default values. We will change some of these for our recipe. Please browse the **Definitions** tab and configure it as follows:

	Mbps	Display Units
17		
18	1280x800	Application Sharing Resolution
19		
20	CIF	Lync 2010 client Max Video Conference Codec on Lync 2010
21		
22	Typical	Lync 2013 client Video codec behavior
23		
24	Typical	Planning Preference - Lync 2010/2013 Codec Behavior
25		
26	70%	Flag in red when WAN BW usage exceeds this value
27		
28	10%	% of Intersite Desktop/App Sharing used in P2P sessions
29		out of total calculated Desktop/App Sharing traffic

You will notice that we have used default values except the following parameters:

- **Display Units**: Change to Mbps for easier readability.
- **Flag in red when WAN BW usage exceeds this value**: Set at 70 percent. This change will be reflected in the readiness diagram that we will review in the later recipe.

Leave everything else in the **Definitions** tab at their default values, that is:

- **Application Sharing Resolution**: This is the assumed resolution for calculations in desktop and application sharing scenarios. You should select the resolution of the desktop screen used by the majority of users.
- **Lync 2010 client Max Video Conference Codec on Lync 2010**: This determines what resolution will be used for video conferences by Lync 2010 clients. In our example, we assume all clients are Lync 2013, so this parameter is irrelevant.
- **Lync 2013 client Video codec behavior**: We are leaving typical as the value. This means we expect the users' video to be based on a static background and moderate moves (typical conferencing scenario), thereby allowing the codec to use less than the maximum bandwidth required.

There's more...

The parameters you set in the definitions tab have a significant impact on the global calculations. It's worth mentioning the following two sections although we are leaving the default values:

- **Traffic from these Lync modalities is included in the Lync traffic**: All set to **Yes** by default, you will need to change these parameters if you are not deploying all Lync communication workloads or you do not want these to be reflected in the calculations. For example, if your deployment does not include Enterprise Voice, you would set the PSTN audio modality to **No**. Note that the parameters configured here affect the global design. If you plan to deploy granularly different features at specific sites, you want to do that through multiple persona definitions.
- **QoS traffic classification for Lync modalities**: The default values are the recommended QoS traffic classes assigned for each modality. If you do not use QoS, these parameters have no effect on calculations.

Defining personas for the network

The **persona** is a fundamental concept in the network readiness calculations; the more accuracy we can come up with while defining it, the closer to reality our estimations will be. A persona represents a category of users with a defined set of usage behavior and feature set for each of the following Lync communication modalities:

- IM/Presence
- Peer-to-peer audio
- Peer-to-peer video
- Conference audio
- Conference video
- Desktop sharing
- PSTN audio

Persona profiles can be configured through a set of predefined standard usages (high, medium, and low), which are based on the following concurrency percentages:

Usage Models Definitions (Max Concurrency at Peak Time per Modality)				
usage	none	low	medium	high
IM/Presence	0%	65.00%	80.00%	90.00%
Presence				
P2P audio	0%	0.50%	1.50%	2.50%
P2P video	0%	0.10%	0.30%	0.50%
conf audio	0%	1.00%	3.00%	5.00%
conf video	0%	0.10%	0.50%	1.00%
desktop share	0%	0.50%	1.00%	1.50%
PSTN audio	0%	5.00%	10.00%	15.00%

What do these percentages mean? Let's take the medium usage for PSTN audio as an example. This value is set at 10 percent. This means (assuming 100 users with such persona profiles in a given site) that the tool will calculate for 10 simultaneous active PSTN calls at any given time.

Persona definitions can be fully customized for each modality. In our test scenario, we will be using standard usage definitions.

How to do it...

Browse to the persona sheet in the Bandwidth Calculator tool. You will notice some predefined persona definitions. Delete them and replace them with the following:

Persona Definitions		2	3	4	5	6	7	8
	Client	Deployment	IM/Presence	P2P audio	P2P video	conf audio	conf video
Standard User	Lync 2013	Onpremise	medium	medium	medium	medium	medium
Heavy conference user	Lync 2013	Onpremise	low	low	low	high	high
Mobile user	Lync 2013	Onpremise	medium	medium	medium	medium	medium

9	10	11	12	13	14	15
		Lync 2010	Remote	Lync 2013	Lync 2013	Lync 2013 users behavior for	Lync 2013
desktop share	PSTN audio	RTV_Type	Users	Stereo Audio	Video Quality	P2P video window	MultiView usage
medium	medium	CIF	10%	0%	Medium	Typical	Typical
high	low	CIF	10%	2%	Best	Typical	Typical
low	medium	CIF	80%	0%	Medium	Typical	Typical

Designing a Lync Solution – The Overlooked Aspects

 [The image is split into two parts for better readability.]

Let's review the rationale for our persona definitions:

- **Standard user**: We have set **medium** or **typical** usage for all modalities. This persona has a low (10 percent) **remote users** percentage. This parameter indicates how many users are to be considered remote. When a user is remote, they will connect through the Lync edge servers from the Internet and the network impact will differ.

- **Heavy conference user**: We have set **low** usages for IM, presence, and peer-to-peer modalities, and **high** usages for conferencing modalities. We have also set the **Lync 2013 Stereo Audio modality** to **2 percent** (the valid range is from 0 percent to 3 percent). Stereo audio is used by Lync Room System devices. We have also set **video quality** to **best**. Multiparty HD video in Lync is supported but has a greater impact on network bandwidth.

- **Mobile user**: Parameters are substantially similar to those assigned to the standard user persona, with the exception of the **remote users** parameters set at **80 percent**. This means we expect a very high percentage of users for the site that is not connected to the WAN network, instead being Internet-based users connecting through the Edge server.

 [The Lync mobile client is always an external client, and it connects via the reverse proxy and leverages the Edge servers for media traffic. This is applicable irrespective of whether the mobile client connects through an internal corporate Wi-Fi network.]

Defining sites for the network

The next and last step before we can perform the actual calculations is to populate the tool with the site-based parameters we identified earlier.

How to do it...

Browse the **Sites** sheet and populate the relevant fields as follows:

			Site definitions				WAN link Info				Internet link		CAC
ID	Site Name	Total Users in Site	Central Site Providing User Services	Internet Site	Number of Sites like this	Local PSTN Breakout?	Low delay WAN ? (Default=No)	WAN Link Speed	Input Units	WAN Link BW Allocated to RTC Traffic	Input Units	Internet Link Speed	Input Units	Restrict Audio Codecs Using CAC
Central Sites														
1	Datacenter	0		Datacenter	1	Yes	No	1	Gbps		Mbps		Mbps	No
2		0			1	Yes	No						Mbps	
3		0			1	Yes	No						Mbps	
4		0			1	Yes	No						Mbps	
5		0			1	Yes	No						Mbps	
6		0			1	Yes	No						Mbps	
7		0			1	Yes	No						Mbps	
8		0			1	Yes	No						Mbps	
9		0			1	Yes	No						Mbps	
10		0			1	Yes	No						Mbps	
Branch Sites														
1	New York	0	Datacenter	Datacenter	1	No	Yes	100	Mbps	20	Mbps			Yes
2	Chicago	0	Datacenter	Datacenter	1	No	Yes	100	Mbps	20	Mbps			Yes
3	Dallas	0	Datacenter	Datacenter	1	No	Yes	20	Mbps	4	Mbps			Yes
4	Denver	0	Datacenter	Datacenter	1	No	Yes	50	Mbps	10	Mbps			Yes
5	Miami	0	Datacenter	Datacenter	1	No	Yes	10	Mbps	2	Mbps			Yes
6	Reno	0	Datacenter	Datacenter	1	No	Yes	4	Mbps	1	Mbps			Yes
7	LA	0	Datacenter	Datacenter	1	No	Yes	4	Mbps	1	Mbps			Yes

For the **Central Sites** section:

- Add the **Datacenter** site.
- Set the **Local PSTN breakout** value to **Yes**. This tells the tool that there is a PSTN gateway in the data center site (later in our configuration, we will set the same parameters for the branch sites to **No**, and this will indicate that our PSTN gateway is centralized in the data center).
- Set the **WAN Link speed** and **WAN link BW allocated to RTC traffic** options as per the table in the previous recipe.
- For the **Personas** option, we have assigned no user. This tells the tool that it is a pure datacenter site.

For the **Branch Sites** section:

- Add all identified sites
- Set the **Central** site that provides the user services parameter to the data center. This tells the tool that all the branch site users are hosted in the Lync deployment in the datacenter site.
- Leave the **Number of sites like this** value as **1** (default). This option comes in handy when we have numerous sites with identical values.

- Set the **Local PSTN breakout** value to **No** for all branch sites. This tells the tool that the PSTN gateway in the datacenter's site will be used (if you had a local PSTN gateway on a site, and you set it to **Yes**, the tool would assume that the calls for that site are routed through the local gateway, and therefore, they would not use the WAN link).

- Set the **Low delay WAN for all branch sites** value to **Yes** (the default is **No**). A **No** value is applied if we know that the latency will be 25 ms or more. This will instruct the tool to consider different codecs (which have a different bandwidth requirement) or introduce FEC (forward error correction) to account for potentially less-than-optimal network conditions at the cost of incremental bandwidth usage for FEC redundancy data.

- Set the **WAN Link speed** and **WAN link BW allocated to RTC traffic** options for all branch sites as per the table in the previous recipe.

- Finally, set the **Restrict Audio codecs using CAC** option to **Yes**. This instructs the tool to change the codecs used in calculations, assuming CAC is restricting the usage to less bandwidth-intensive codecs with the purpose of allowing for more sessions (for example, without CAC, the calculator will consider the RTAudio Wideband codec for P2P calls using 45 Kbps for typical usage, whereas with CAC, the tool will calculate based on the RTAudio Narrowband codec typically using 27.8 Kbps).

- For the **Persona** option, we have assigned no user. This tells the tool that it is a pure datacenter site.

Next, let's populate the **Persona** fields in the **Branch Sites** section, as shown in the following image:

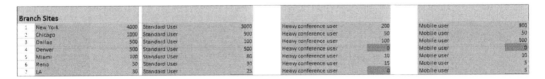

This way, we have assigned a given number of users for each persona and for each site to sum up to the total user count in our initial table.

The tool is now ready to perform the calculations.

See also

For an overview of codecs used by Lync and the typical bandwidth usage for each, please refer to http://technet.microsoft.com/en-gb/library/jj688118.aspx. You can also refer to the Codecs sheet in the Bandwidth Calculator tool and even change them (although we do not recommend it unless you know exactly what you are doing).

Chapter 6

Network readiness – reviewing and analyzing results

In this recipe, we will learn how to analyze and interpret the results, and determine what sites are Lync-ready.

How to do it...

Browse to the **Graphical results** sheet and click on the **Update tables and graphs** green button.

 Remember to do this each time you change any value in the tool.

We will be presented with summary site tables and two diagrams, which will give immediate visual indication of readiness, as shown in the following image (note that we have split the sections for better readability):

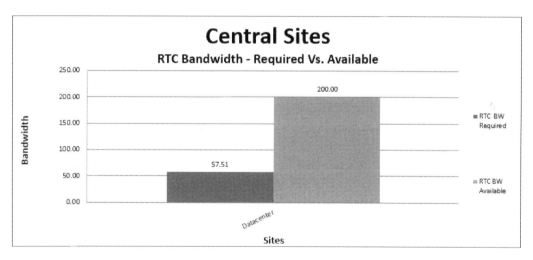

The first picture to look at is the one for **Central Sites**. The light bar in the preceding graph indicates our total RTC bandwidth availability for RTC (200 mbps), and the dark bar is the calculated amount of bandwidth required (57.51 mbps) to support Lync traffic for all branch sites associated with the **Datacenter** central site. The good news here is that we can safely assume our WAN capacity to the datacenter is more than enough.

The following graph provides a detailed breakup of the available bandwidth versus the bandwidth required for each branch office.

The **RTC bandwidth available** value is the one we have earlier predefined as reserved for the Lync RTC traffic.

For example, New York's total WAN capacity is 100 mbps, with 20 mbps reserved for RTC, and the latter is the value reflected in the next image:

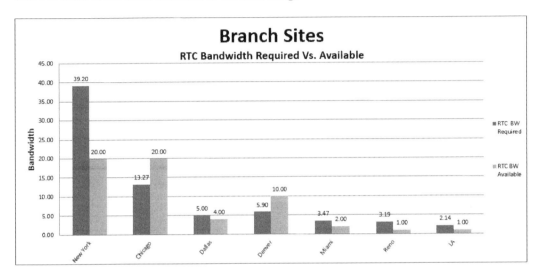

Without digging into the actual numbers for now, we can get an immediate overview through the bar diagram and make the following statements:

- For New York, the required bandwidth is far more than the available one (20 mbps is available, but almost double of what is available is required)
- Miami, Reno, and LA also require considerably more bandwidth than the available one
- Chicago and Denver have considerably more bandwidth availability than required and can be safely regarded as Lync-ready
- Dallas requires more bandwidth than the available one, but there is room for adjustments, which may turn this site into a Lync-ready one.

Now, browse to the **Deployment readiness** tab and review the pie diagrams; these contain a summary of the number of sites that require a WAN upgrade or perhaps usage adjustments before they can be considered Lync-ready.

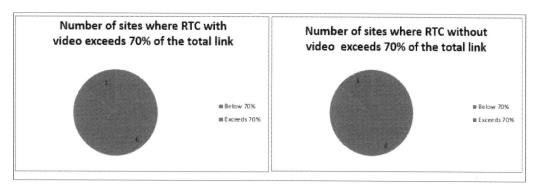

What useful information can we retrieve from the preceding images? Remember when we configured the **Flag in red when WAN BW usage exceeds this value** to 70 percent? This is where the parameter is reflected. It works as a "warning level" that tells us how many sites go beyond 70 percent WAN usage for the total link capacity. Be careful about this point. The takeaway from this output is that if we could reserve a greater amount of WAN bandwidth for RTC, possibly up to 70 percent, *this would result in only one site, exceeding this threshold*.

 In a real-world scenario, you will want to set this threshold to a level that you know will not affect your existing line-of-business applications that already make use of your WAN. Acquiring real-time network usage statistics from your WAN carrier or through specialized network monitoring software is paramount to determine an acceptable threshold.

The following pie diagram takes into account the overall network readiness based on the bandwidth we originally allocated in the calculator (20 percent):

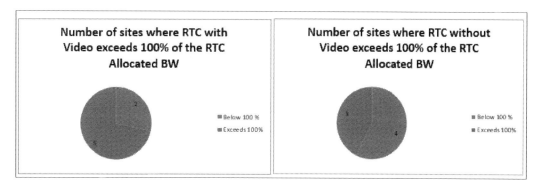

Designing a Lync Solution – The Overlooked Aspects

As shown in the previous diagram, we are provided with two outputs, one with a video and one without. The takeaway from this is as follows:

- If we do consider the video, out of seven sites, two meet network compliance, while five do not.
- If we do not consider the video, out of seven sites, four meet network compliance, while three do not. This provides a potentially useful indication that a tune-up in persona definitions and/or more strict control over allowing videos for some users might turn the situation for some sites and make them Lync-ready without a WAN bandwidth upgrade.

If we need to further break down to actual values, another useful resource is the **sites** tab. Scroll down and then sideward until the following is displayed on your screen:

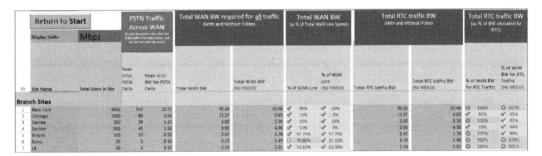

This table reveals the following:

- **Reno** is the only site (that scores 79.80 percent) that is exceeding 70 percent threshold over the total WAN link bandwidth. Provided 70 percent is our non-negotiable hard limit, this confirms that the site requires a WAN upgrade for Lync.
- **Miami** scores 89 percent total RTC bandwidth with the video and 174 percent total RTC bandwidth without video. This is a definite indication that 20 percent reserved bandwidth gets us very close to threshold even without the video, and largely over the limit with video. However, if we consider the maximum nominal Miami WAN link, it only scores 34.74 percent; in other words, doubling-up the reserved RTC bandwidth would render the site RTC-ready.

Chapter 6

There's more...

You might have noticed that the tool allows for great flexibility in setting up and customizing several parameters for each site.

It is important to notice that values might have to be adjusted in specific circumstances, and ultimately, you are leveraging statistical usage models (personas) that cannot guarantee a perfect match with the real world. An exact calculation is only revealed through usage; furthermore, network usage can continuously vary depending on the user's evolution, trends (video adoption is a slowly but constantly growing one), and evolving business objectives.

In these aspects, the tool provides an excellent overview, and ultimately, it is an essential resource to highlight likely network bottlenecks that need to be addressed before deployment as an essential milestone in every successful Lync deployment project.

See also

- The **Lync 2013 and 2013 bandwidth calculator tool** download package includes a user guide that covers advanced scenarios. It is available at http://www.microsoft.com/en-us/download/details.aspx?id=19011.

7
Lync 2013 in a Resource Forest

In this chapter, we will cover the following recipes:

- Introduction
- Planning a resource forest
- Using Exchange Online for Lync resource forests
- Configuring FIM in a Lync resource forest
- Synchronizing forests with FIM
- Deploying Azure Active Directory Synchronization services (AAD Sync) in a Lync resource forest
- AAD Sync synchronization service and rules

Introduction

Deploying Lync has a cost both on time and money. In addition, the more features, availability, and quality we deliver, the more resources will be required. It makes sense for a company to dampen costs, deploying Lync Server 2013 in a single domain/forest and making Lync services available to users whose accounts reside on separate forests outside the single corporate boundary. There are two different topologies we are able to deploy to achieve the previously mentioned result: **Lync in a Central Forest** or **Lync in a Resource Forest**. In both scenarios, we have a two-way forest trust between the forest-hosting Lync and the external forests where the user accounts reside. In the first topology, the forest that contains our Lync deployment is called the central forest.

There are active user accounts in all the forests, including the central forest, and the ones that have no Lync Server available are called **User Forests**. A resource forest topology is similar, and we have the Lync Server installed in a single resource forest. However, our active user accounts are located only in the user forests. We can easily understand this by looking at the following image:

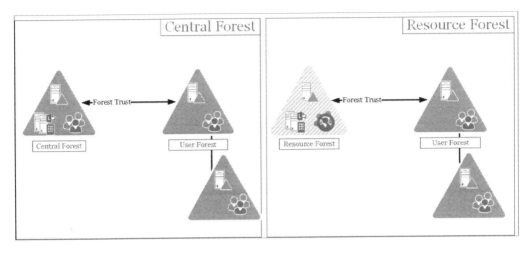

Planning a resource forest

The resource forest hosts disabled user accounts (used to enable access to Lync services). It is required that the **msRTCSIP-OriginatorSID** attribute for the disabled accounts maps to the **ObjectSID** of the account in the user forest. The resource forest topology has a higher level of isolation between Lync and the users' forest, but the increased security also implies increased administrative complexity.

This can be summarized as follows:

- We have to select a topology (central or resource forest)
- We have to configure forest trusts to enable users' authentication into the resource forest

Also, if we have not talked about this aspect before, we have to add another point to the list.

- We have to select a tool (automatic or manual) to create a match between the information required in the resource forest and the ones available in the user forest

If we have already deployed Exchange in our resource forest, we can take advantage of the attribute `msExchMasterAccountSid` that works as the previously mentioned `msRTCSIP-OriginatorSID`.

The `msExchMasterAccountSid` attribute is used to grant Exchange services to the user forests via a disabled account. As we shall see, `msExchMasterAccountSid` joins with the ObjectSID. Exchange has a practical wizard that helps to map Linked Mailboxes (mailboxes that are accessed by users in a separate, trusted forest) to user accounts in their originating forest. The same operation is available using the Exchange Management Shell. We will be able to use `msExchMasterAccountSid` to populate the `msRTCSIP-OriginatorSID` attribute required for Lync. The outline of the laboratory that we will use is shown in the following image:

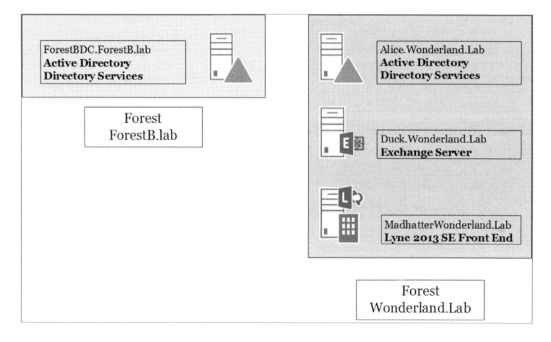

This is the recipe we will see for planning the steps required to deploy Lync in a resource forest that already contains an Exchange deployment.

Lync 2013 in a Resource Forest

Getting ready

Our scenario is based on the assumption that a working Exchange 2013 deployment is already available in the resource forest. In addition, we will use the Lync 2013 Resource Kit (http://www.microsoft.com/en-us/download/details.aspx?id=36821). In a folder inside the installation path of the resource kit, `LCSSync`, we have a script called `sidmap.wsf`. This utility takes the existing SID information from the `mxEXCHMasterAccountSID` attribute (populated by Exchange when we have defined a linked mailbox) and uses this value to fill the `msRTCSIP-OrginatorSID` attribute, as shown in the following screenshot:

How to do it...

1. From **Exchange Admin Center (EAC)** (accessible through the URL at `https://<CASServerName>/ecp`), in the **Recipients** menu, open **Mailboxes**.

2. Click on **New** and select **Linked mailbox**. Follow the wizard to select a trusted forest or domain, as shown in the next image:

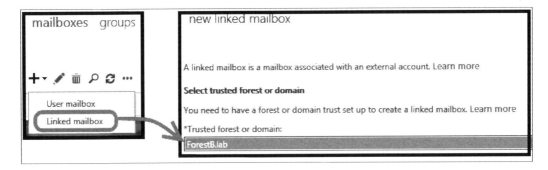

3. We will be required to enter the administrative credentials for the trusted domain. In the next screen, we must identify the linked master account, as shown in the next screenshot:

Chapter 7

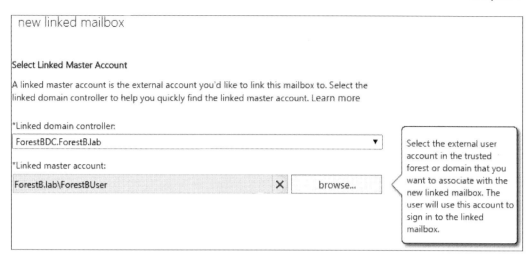

4. The wizard requires information about the new disabled account, which will be defined in the resource forest, as shown in the following screenshot:

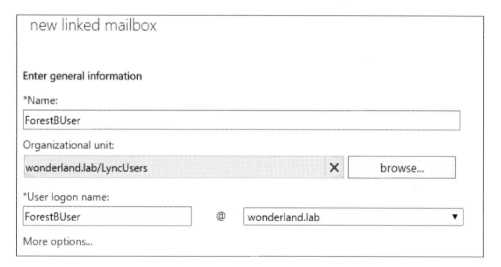

5. Click on **Finish**.

6. Using **ADSIEDIT**, we are able to verify **objectSid** of the user in the user forest and compare it with the `msExchMasterAccountSid` attribute, as shown in the following screenshot:

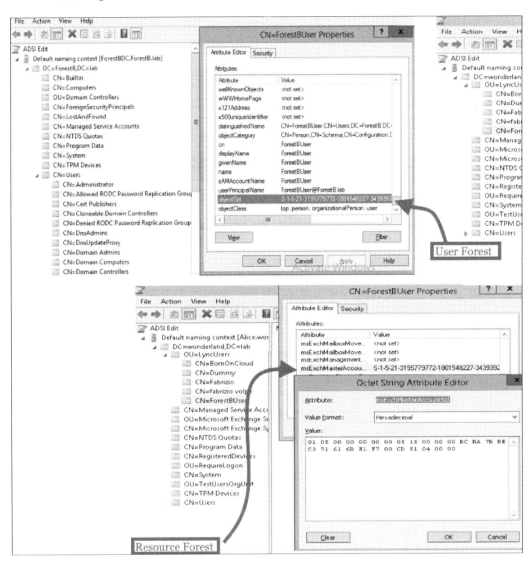

—— *Chapter 7*

7. Now it is possible to enable to Lync the disabled account that resides in the resource forest to Lync, as shown in the following screenshot:

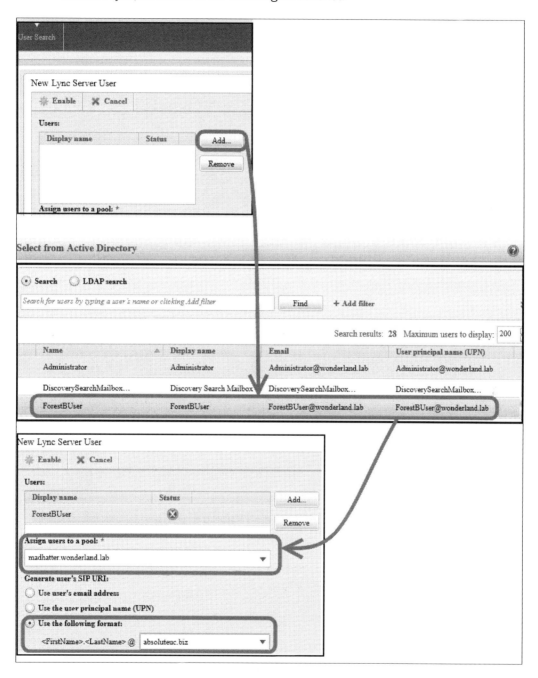

Lync 2013 in a Resource Forest

8. We have to synchronize the `mxEXCHMasterAccountSID` attribute with the `msRTCSIP-OrginatorSID` attribute for the user account that we have SIP-enabled in the previous step. We will use the `sidmap.swf` script. In our example, the users will be located in the **LyncEnabled** OU (**OU=LyncEnabled,DC=wonderland,DC=lab**). The path to the **sidmap.wsf** script is `C:\Program Files\Microsoft Lync Server 2013\ResKit\LcsSync`.

9. Open a command prompt as an administrator and launch the following commands:

    `cd "C:\Program Files\Microsoft Lync Server 2013\ResKit\LcsSync"`

    `wscript //h:cscript`

    `sidmap.wsf /OU:"OU=LyncEnabled,DC=wonderland,DC=lab" /logfile:c:\sipmap.txt`

    We can verify the result of the script both on the command prompt and on the `sipmap.txt` file, as shown in the following screenshot:

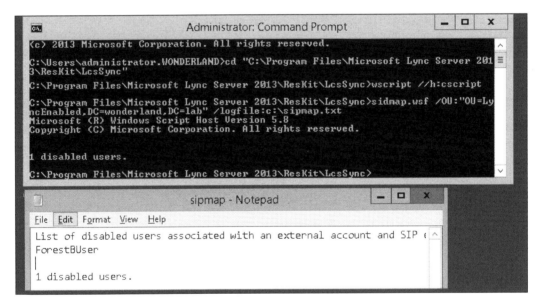

10. Using **ADSIEDIT**, we are able to verify whether the two attributes are aligned, as shown in the following screenshot:

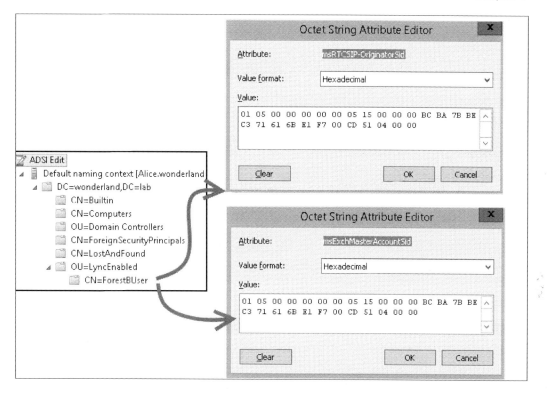

There's more...

Users that are not mail-enabled have no data in the msExchMasterAccountSid attribute. We will have to manually map the active user's SID to the msRTCSIP-OriginatorSID attribute of the disabled account every time there is no linked mailbox.

The following code is an example of a script we can use to align msExchMasterAccountSid to msRTCSIP-OriginatorSID. We should save the text inside a .ps1 file (for example, enableUC.ps1) and launch it using the following cmdlet .\enableUC.ps1 -name "username" -pool "LyncPoolName":

```
Param([String] $Name, [String] $Pool)
$user = Get-ADuser $name -Properties msExchMasterAccountSid,msRTCSIP-OriginatorSid,proxyAddresses
if (Get-CsUser $user.UserPrincipalName 2> $null)
{
```

```
    echo "User Already Enabled"
    Get-CsUser $user.UserPrincipalName
    exit

}
foreach ($smtp in $user.proxyAddresses)
{
   if ($smtp -clike "SMTP:*")
   {
        $sip = $smtp.ToString()
        $sip = $sip.Replace("SMTP", "sip")
        $sip
   }
}

if ($user.msExchMasterAccountSid -notlike $null)
{
    Set-ADUser -Identity $user -Add @{"msRTCSIP-OriginatorSid" =
$user.msExchMasterAccountSid.Value.ToString() }
    enable-CsUser -Identity $user.UserPrincipalName -RegistrarPool
$Pool -SipAddress $sip
    echo "User Enabled to Lync"
    Get-CsUser $user.UserPrincipalName
    exit
}
```

See also

Shawn Cathcart wrote a couple of handy scripts to automate the enabling of the cross-forest Lync user and the SID attributes' verification. They are available in the *User Enabling in a Lync Resource Forest Deployment* post at http://thecathcart.blogspot.it/2012/07/user-enabling-in-lync-resource-forest.html.

In the previously mentioned post, there is also important information about Lync RBAC in a resource forest scenario. The forest preparation step of the Lync 2013 installation creates the **Role-Based Access Control** (**RBAC**) groups (we have mentioned them in the *Controlling administrative rights with RBAC and custom cmdlets* recipe of *Chapter 1, Lync 2013 Security*). RBAC makes it easier to delegate administrative tasks that grant limited permissions and tools to the user that receives the delegation. In a resource forest scenario, the forest preparation is executed only in the forest that hosts the Lync services. Lync RBAC Roles are associated with Universal Security groups in the resource forest. Only local security groups can take members and groups from a trusted forest, but we are not able to add them in a universal security group. The creation of a domain local group, inserting a user forest's account, and adding the local group in the RTCUniversalReadOnlyAdmins group is shown in the following screenshot:

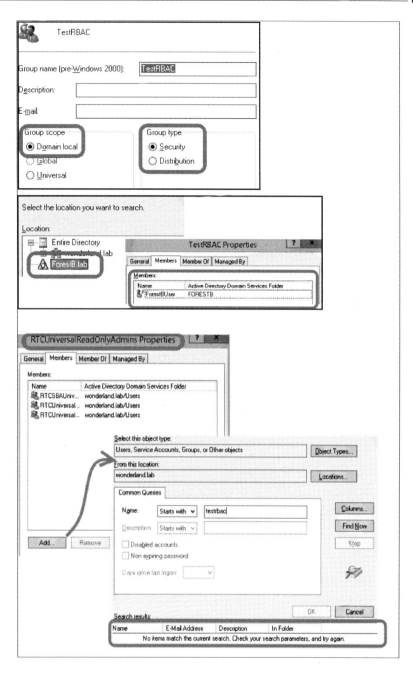

So, while there are some ways to overcome this problem with Exchange in a resource forest (see the *Understanding multiple-forest permissions* post at http://technet.microsoft.com/en-us/library/dd298099(v=exchg.150).aspx), the only way to administer Lync is to have an account in the same forest as the servers that are deployed.

Using Exchange Online for a Lync resource forest

What we have seen in the previous section applies (for a large part) also to companies that have moved Lync (and Exchange) to the cloud but still require a resource forest configuration. Microsoft has published a document, *Deploying Lync in a Multi-Forest Architecture (Partner Hosted Lync with Exchange Hybrid)*, at `http://www.microsoft.com/en-us/download/details.aspx?id=44276`, which outlines how to realize different multi-forest hybrid scenarios.

The main issue in a hybrid resource forest is related to companies requiring Enterprise Voice for Lync which is not available on Office 365. In a scenario like the previously mentioned one, our resource forest will be split between two different hosting providers (on-premises and cloud). The two hosts have no trust relationship, and this makes the deployment more complex. Exchange deployment can be online or hybrid (with an on-premises component), and depending on the decision we take about our Exchange services, our resource forest scenario will change. We will see some of the steps related to the resource forest deployment including only Exchange Online services. The Lync resource forest will contain disabled accounts.

Exchange Online will use active accounts related to our domain accounts that are enabled to mail services. That's a DirSync-based configuration, like the one we have seen in *Chapter 2, Lync 2013 Authentication*, with an identity management solution such as Forefront Identity Manager, which we will see in the following sections. The following schema is a high-level overview of our scenario (our user forest will have a public name of `ForestB.com`):

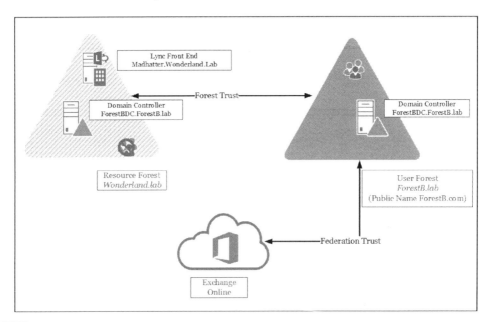

How to do it...

1. **Internal DNS**: While the DNS records for the resource forest and for Exchange Online are the ones we use in a standard deployment, Lync records on our user forest will use DNS records that point to the resource forest (with a split brain configuration). For Exchange, there are two required records that point to Office 365:

CNAME	`autodiscover.outlook.com`
CNAME	`msoid.ForestB.com`

2. **Public DNS**: Public records for the ForestB.com Lync deployment will be the usual ones that point to the resource forest. For Exchange, we have the following records to configure (which point to Office 365):

CNAME	`autodiscover.outlook.com`
CNAME	`msoid.ForestB.com`
A	`mail.ForestB.com`
MX	`ForestB.mail.eo.outlook.com`
TXT	`ForestB.com`

3. For the purpose of securing access to the internal resources, Lync is usually based on certificates from an internal CA. We have to import the certificate of the issuing CA (and eventually the whole chain) in our user forest. The Microsoft document that we mentioned in the introduction of this paragraph offers a solution based on **certutil**. Another solution that only uses the Group Policy Management is outlined in the TechNet post Manage Trusted Root Certificates at `http://technet.microsoft.com/en-us/library/cc754841.aspx`.

   - Export the CA certificate from the resource forest (in our scenario, the CA is `alice.wonderland.lab`) and copy it to a domain controller in the user forest.
   - Launch **Group Policy Management**, edit the **Default Domain Policy** option, and select **Computer Configuration**, **Windows Settings**, and **Security Settings**.
   - Click on **Public Key Policies**.

- Right-click on the **Trusted Root Certification Authorities** store and import the CA certificates, as shown in the following screenshot:

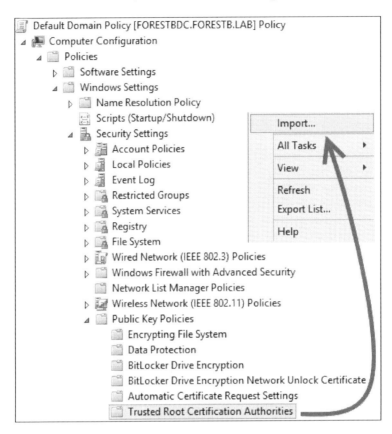

4. Configure Federation for SSO and Directory Synchronization for Exchange Online (see Chapter 2, Lync 2013 Authentication and then see the Authenticating with online services using DirSync recipe).

5. Lync and Exchange Online Identity management require a synchronization service. We will talk about FIM for the synchronization of the Lync resource forest with the user forest in the following sections. For Exchange Online, we are able to use the **Forefront Identity Manager Connector for Windows Azure Active Directory** or the **Azure Active Directory Synchronization Services** (**AAD Sync**). We will look at the configuration and use of AAD Sync later in this chapter.

6. Configure the Lync resource forest like a standard Lync deployment. Additional configuration requirements are the DNS records in the user forest and (if deployed) integration with Exchange Online UM.

7. For the deployment of Exchange Online, for the most part, we will follow standard procedures. It is important, however, to map the disabled accounts in the Lync resource forest to Exchange Online users. Attribute mapping for on-premises to the cloud could be complex, but if we are using AAD Sync, all the configuration is automatic because AAD Sync setup will detect our existing deployment of Lync and Exchange, and create dedicated rules and connectors (refer to the sections about AAD Sync in this chapter).

Configuring FIM in a Lync resource forest

As we have seen in the previous recipe, a Lync resource forest requires a continuous synchronization between the accounts in the users' forests and the disabled accounts that must be provided in the forest in which Lync services exist. A directory synchronization product, such as **Microsoft Forefront Identity Manager (FIM) 2010 R2**, is a useful solution to transmit modifications from the user forests to the resource forest (for example, creating or deleting an account in the former one will automatically create or delete the disabled account in the latter). In the following schema, we can see a possible outline for a resource forest deployment with FIM (Identity Manager is not joined to the user and resource forest to show there is no need to insert it inside the existing forests):

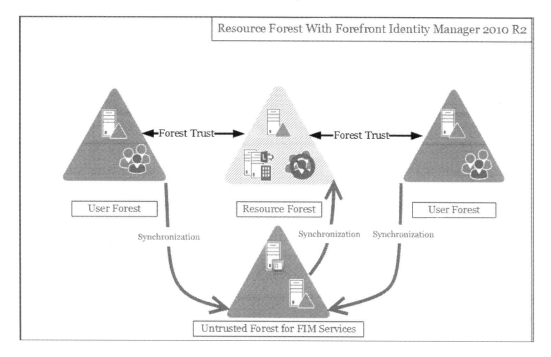

This section will be dedicated to the configuration of FIM in a resource forest scenario.

Lync 2013 in a Resource Forest

Getting ready

To install FIM 2010 R2, it is required to have a SQL server installation available. We can co-host the database on the FIM server or have a dedicated database server. In our scenario, we have a dedicated forest **FIMDomain.Lab**. The server that will host both FIM and the SQL 2012 database for it is **FIM.FIMDomain.Lab**. In the following screenshot, we can see a high-level overview of the deployment:

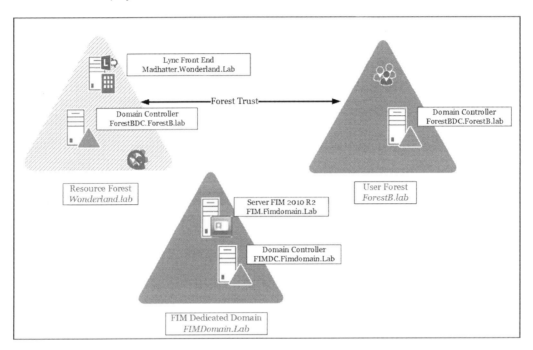

FIM installation will require a service account, which we will call **FIMService**. It requires no special permissions.

> FIM is a really complex software to deploy, and it has a heavy impact on the Active Directory functionalities. The information in this recipe is really basic, so it is a recommended approach to study and plan an FIM deployment in a test environment first.

How to do it...

1. The first step is to install **SQL Database Engine Services** and **Reporting Services – Native**. The selection is shown in the following screenshot:

Chapter 7

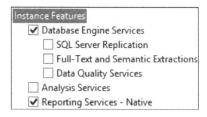

We will not see how to install SQL Server 2012; however, there are many dedicated resources, such as the TechNet post *Installation How-to Topics* at http://msdn.microsoft.com/en-us/library/cc281837(v=sql.110).aspx.

2. We will install the Synchronization Service. We are able to launch the setup from the installation media (for example, D:) using D:\Synchronization Service\Synchronization Service.msi. The installer will open the **Welcome** page. Select **Next**.

3. Then, select **I accept the terms in the License Agreement** and click on **Next**.

4. We have to select an installation path and then click on **Next**.

5. The **Service Database Connection** screen will ask for information about our SQL database. If we co-hosted SQL on the FIM server (installing it with the default settings), we will be able to just click on **Next**. If we have deployed a separate SQL server, we have to insert the connection information, as shown in the following screenshot:

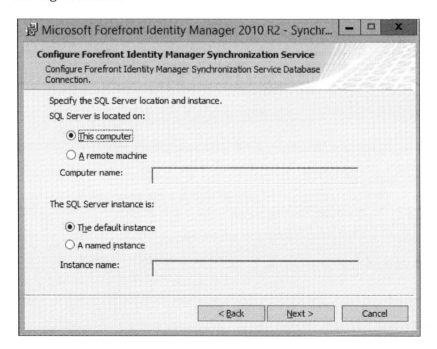

6. FIM installation will require the information about the service account (FIMService). We have to insert the username, password, and domain, and then click on **Next**.

7. As we can see in the following screenshot, FIM will automatically create new security groups. We can just leave the defaults and click on **Next**:

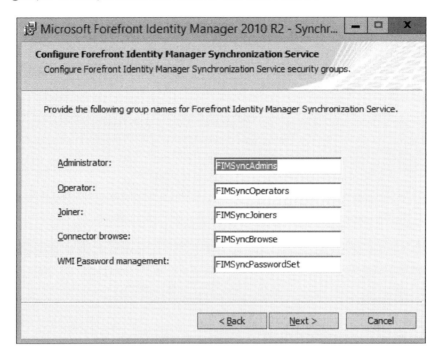

8. The next step will propose the creation of firewall rules to allow access from the clients to the FIM server. Depending on our server configuration, we can decide if we want to flag it or not:

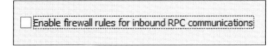

9. Select **Next** and then **Install** in the following screen. During the installation, a backup of the encryption file (with a `.bin` extension) will be created. We should keep it in a safe place outside of the server.

10. At the end of the installation process, we have to select **Finish**. A logoff and logon will be required to activate the membership of the FIM security groups.

Chapter 7

11. We have to copy the contents of the `LcsSync` folder from a Lync 2013 Resource Kit installation to the FIM server. In our example, the path of the Resource Kit is `C:\Program Files\Microsoft Lync Server 2013\ResKit`. The path where we have to copy the content of the previously mentioned folder on the FIM server is the `C:\Program Files\Microsoft Forefront Identity Manager\2010\Synchronization Service\Extensions` folder, and can be seen in the following screenshot:

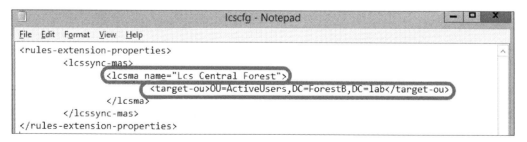

12. In our user's forest, we have an OU named **ActiveUsers** (**OU=ActiveUsers,DC=Forest B,DC=lab**), which contains the accounts we will import in the resource forest.

13. On the FIM server, we have to edit the `lcscfg.xml` file in the previously mentioned `Extensions` folder. The `target-OU` value must be equal to the value of the OU that contains the active accounts in the user forest (in our example, **OU=ActiveUsers,DC=ForestB,DC=lab**). The `lcsa` name parameter will be used during the configuration of the management agent. In the following screenshot, we can see the edited file:

```
<rules-extension-properties>
        <lcssync-mas>
                <lcsma name="Lcs Central Forest">
                        <target-ou>OU=ActiveUsers,DC=ForestB,DC=lab</target-ou>
                </lcsma>
        </lcssync-mas>
</rules-extension-properties>
```

14. Launch the **Synchronization Service Manager** window, select **Metaverse Designer**, click on **Actions**, and select **Import Metaverse Schema**, as shown in the following screenshot:

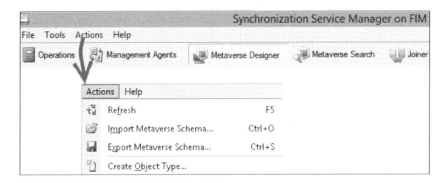

15. Select the `Lcsmvschema.xml` file from `C:\Program Files\Microsoft Forefront Identity Manager\2010\Synchronization Service\Extensions`, as we can see in the following screenshot:

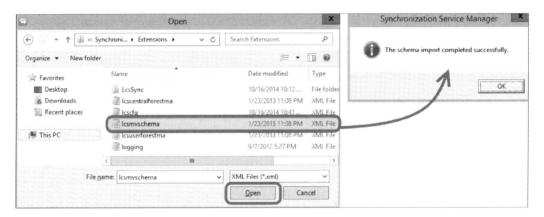

16. Select **Tools**, click on **Options**, and select **Enable metaverse rules extension flag**. Click on **Browse** and select **lcssync.dll**. Click on **OK** as shown in the following screenshot:

Chapter 7

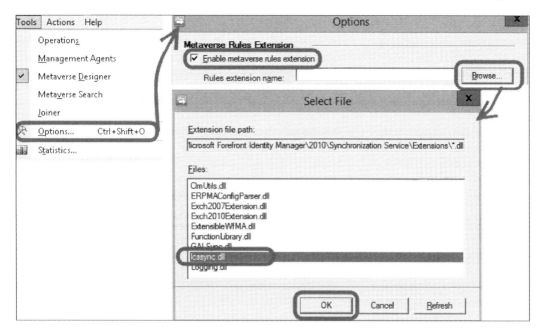

17. In the **Options** screen, select the **Enable Provisioning Rules Extension** option and then click on **OK**. The option is shown in the following screenshot:

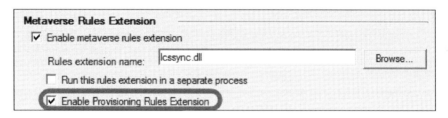

How it works...

FIM is based on two base components: the metaverse and the connector space. The metaverse is a MetaDirectory (a system to collect, aggregate, and store data from various directories and data sources, such as Active Directory). The metaverse is stored in five SQL tables where information is organized using a schema (details about the FIM schema are available in *Understanding Custom Resource and Attribute Management* at http://technet.microsoft.com/en-us/library/ff519007(v=ws.10).aspx). FIM uses management agents to update data in the metaverse and in the data sources. The other component, connector space, is a temporary storage area for entities (objects). Data is modified, deleted, or added in the connector space before flowing to the metaverse or the data sources. A modified "shadow copy" of the data source is stored here by management agents.

Synchronizing forests with FIM

In the previous section, we discussed the management agents used by FIM. The agents' actions are activated using a series of operations to import, synchronize, and export information from the data sources (our Directory Services) to the metaverse and vice versa. In the previous steps, our FIM deployment has been configured to support Lync in a resource forest configuration, so now we have to enable the agents and the operations. Import the `lcscentralforestma.xml` file and click on **OK**.

> The `Fabrikam.com` and `NWTraders.com` partitions are the default partition names used when we import the FIM management agents. We are going to keep the names as they are, configuring `Fabrikam.com` with the information of the resource forest, and `NWTraders.com` with the information regarding the user forest.

How to do it...

1. We will start with the management agent for the resource forest (`wonderland.lab`). We have to go to the **Management Agents** screen, select **Actions** and **Import Management Agent**, and then select the `lcscentralforestma xml` file, as shown in the following screenshot:

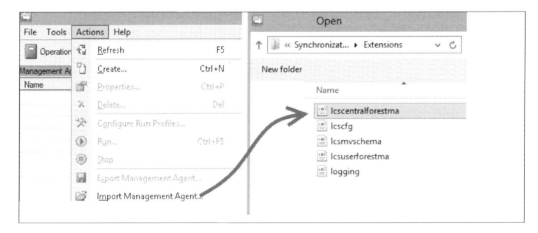

2. The first page of the management agent configuration will contain the name that we have inserted in the `lcscfg.xml` file during the previous section. Select **Next**.

3. Insert the account information in order to connect to the resource forest and then click on **Next**.

4. Click on **Match**, and select **DC=FABRIKAM,DC=COM** and **DC=wonderland,DC=lab**. Click on **Deselect** for all the remaining existing partitions, as shown in the following screenshot:

Updated partition	Operation	Existing partition
	Deselect	CN=Configuration,DC=rtcstore,DC=nttest,DC=m...
	Deselect	DC=DomainDnsZones,DC=wonderland,DC=lab
	Deselect	DC=ForestDnsZones,DC=wonderland,DC=lab
DC=FABRIKAM,DC=COM	Manual Match	DC=wonderland,DC=lab

5. Click on **OK**.

6. The next screen, **Configure Directory Partitions**, offers options to select a preferred domain controller, to modify the connection security, and to define the OU that we will use to import the users (as we are talking about the resource forest, the flow of account information should be inbound only). In our scenario, the OU is called **ActiveUsers**. Flag it as shown in the following screenshot and click on **OK**. Then, select **Next**:

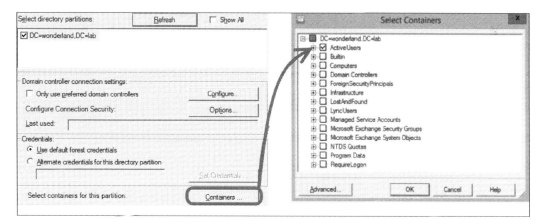

7. The remaining options are already correctly configured based on the configuration we have imported. For example, by looking at the **Select Attributes** screen, you will notice that we have selected all the AD attributes that are relevant for Lync. Also, the **Configure Extensions** options are automatically set to point to the `lcssync.dll` file, which we have previously imported, as shown in the following screenshot:

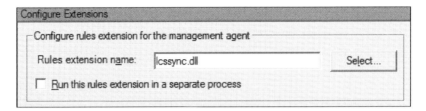

8. Click on **Finish** to complete the definition of the management agent.

9. A similar process is required to define the management agent for the user forest (**ForestB**). We have to go to the **Management Agents** screen, select **Actions** and **Import Management Agent** and then select the `lcsuserforestma` XML file.

10. The name of the management agent will be **ForestB**. As we have seen before, once again insert the account information to access the user forest with the agent. This time, the partition matching rules will be the ones shown in the following screenshot (which match the default partition **NWTraders.com**):

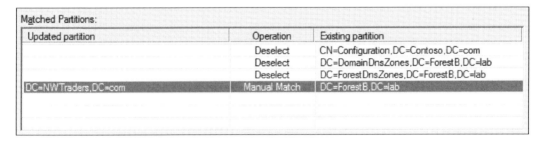

11. The OU that we will use to synchronize is called **ActiveUsers**.

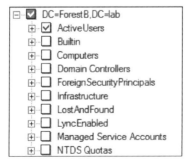

Chapter 7

12. We will move all of the users that we plan to Lync-enable from the user forest to the OU that we previously selected.

13. By right-clicking on the management agents, we are able to run various operations to import or sync data between forests. For instance, the order required to synchronize a resource forest is as follows:

    - Perform a full import on the resource forest management agent
    - Perform a full import on the user forest management agent
    - Perform a full sync on the resource forest management agent
    - Perform a full sync on the user forest management agent
    - Export the resource forest management agent

There's more...

During the configuration of the management agent for the user forest, we might encounter the following error: **msExchUserHoldPolicies of inetOrgPerson is no longer available**.

It is a nonexistent attribute, so we have to remove it from the Metaverse Attributes list as shown in the following screenshot (it is inside **Object Type: inetOrgPerson**):

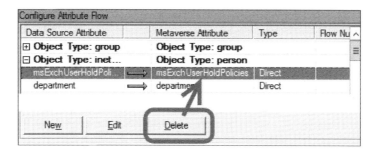

A second similar error will be generated for the user data source. Again, expand the user and delete `msExchUserHoldPolicies`. See the following screenshot:

Deploying Azure Active Directory Synchronization services (AAD Sync) in a Lync resource forest

It is possible to deploy a resource forest using Exchange and Lync in their Online version. For a similar scenario, there is a **Forefront Identity Manager Connector for Windows Azure Active Directory** (http://www.microsoft.com/en-us/download/details.aspx?id=41166). However, in the past few months, Microsoft has published a new tool, **Azure Active Directory Synchronization Services** (**AAD Sync**). Quoting the MSDN site http://msdn.microsoft.com/en-us/library/azure/dn790204.aspx, this new synchronization service allows the user to:

> "Synchronize multi-forest Active Directory environments without needing the full blown features of Forefront Identity Manager 2010 R2".

Right now, this tool is in the general availability stage. In the *How it works...* section of this recipe, we will talk about the AAD Sync working logic. Now, we will see how to deploy it.

Getting ready

We need the installation files for AAD Sync, available at the **Microsoft Azure Active Directory Sync Services** page (http://www.microsoft.com/en-us/download/details.aspx?id=44225). The server that we will dedicate to AAD Sync must be joined to a domain that runs Windows Server 2008 SP2 or higher. It is necessary to deploy an Azure account with an **Active Directory** service, as shown in the following screenshot:

From a security point of view, it is a good practice to create a dedicated global administrator. The Directory Sync must be **Activated** on the user forest's Active Directory, as shown in the following screenshot:

Chapter 7

Our scenario is based on a resource forest (`Wonderland.lab`) with Lync 2013 and Exchange 2013 deployed, a user forest (`ForestB.lab`), and an untrusted third forest (`FIMDomain.lab`). We will use the latter to install AAD Sync.

How to do it...

1. Launching the installer on the **Welcome** screen will require the user to enter the installation path and to accept the license agreement, as shown in the following screenshot:

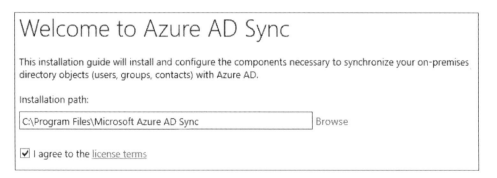

2. Click on **Install**. In the next screen (**Connect to Azure AD**), the username and password of a global administrator for our Active Directory service in Azure are required. Insert the information and then click on **Next**. The screen is the one shown in the following screenshot:

3. The setup process will now require credentials (the domain name, username with the domain or account format, and password) to connect to our domains. In our example, we have the resource forest (wonderland.lab) and the user forest (forest.lab). The configuration is the one we can see in the following screenshot. For every forest, we have to click on **Add Forest** to confirm the information.

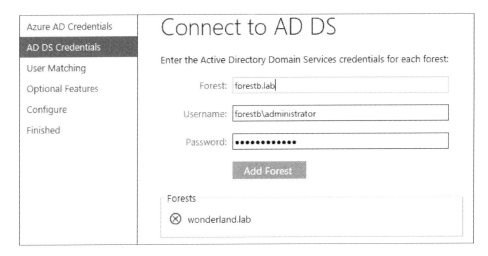

4. Once a forest is added, AADSync will create an initial default configuration based on the services that the forest contains, such as Exchange and Lync.

5. Click on **Next** when there is no more forest to add. The **Uniquely identifying your users** screen requires the user to select the matching attribute to use across the different forests and the matching that we will use on the Azure Active Directory. With Exchange deployed in our resource forest, we will use ObjectSID and mxEXCHMasterAccountSID, while Azure AD will rely on ObjectGUID/ userPrincipalName. The configuration is shown in the following screenshot:

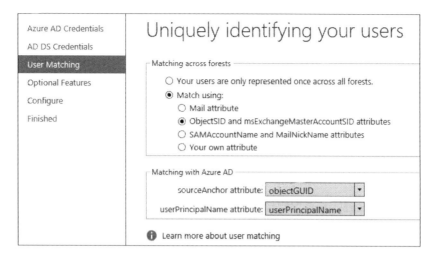

Click on **Next**. The next page, **Optional Features**, does not require any modifications for our scenario. Click on **Next**.

6. The next screen, **Ready to configure**, will show a list of the domains and AD services we are going to connect. Click on **Configure**, as shown in the following screenshot:

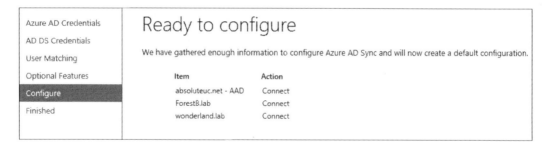

7. The last step is the **Finished** screen. By default, the **Synchronize now** flag will be selected. To manage the sync service, it is required to log off and log on again to apply the **ADSyncAdmins** membership. Click on **Finish**.

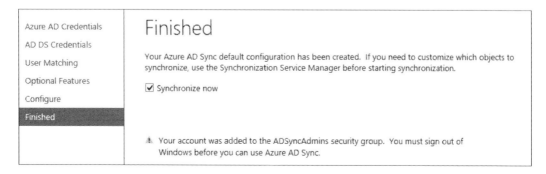

Lync 2013 in a Resource Forest

How it works...

The concepts of metaverse and connector space that we have seen for FIM also apply to AAD Sync. There is no management agent in AAD Sync; all the data is gathered by the server using **connectors** (remote connections to the data sources). As we mentioned before, the flow of information can be inbound or outbound. We have a high-level overview in the following schema:

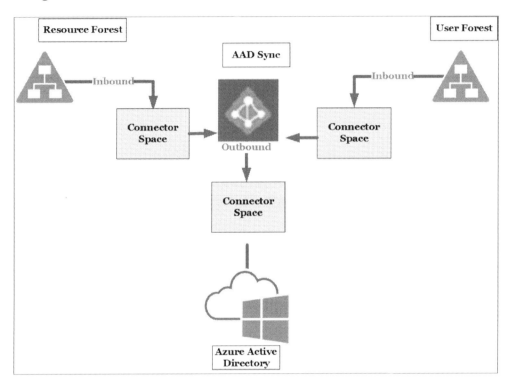

AAD Sync synchronization services and rules

AAD Sync management is performed using two different interfaces, the **Synchronization Service** option and **Synchronization Rules Editor**, as shown in the following screenshot:

Chapter 7

Now, we will see how to manage some aspects of AAD Sync using the previously mentioned administrative tools. The outline of our lab deployment is shown in the following image:

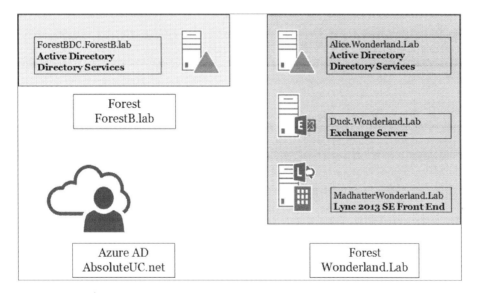

How to do it...

1. We have discussed the metaverse in the previous section. The way information flows inside and outside the metaverse is dictated by the synchronization rules. As we can see in the following screenshot, inside the **Synchronization Rules Editor**, we have a "direction" for every rule (**Inbound** and **Outbound** to the metaverse).

Rule Types	Name	Connector
Inbound	In from AAD - Contact Join	absoluteuc.net - AAD
Outbound	In from AAD - Group Join	absoluteuc.net - AAD
	In from AAD - User Join	absoluteuc.net - AAD

203

Lync 2013 in a Resource Forest

2. The default configuration includes inbound rules from all our domains (the resource forest, `wonderland.lab`, the user forest **ForestB.lab**, and the Azure Active Directory `absoluteuc.net`) while the outbound rules are only for the Azure Directory.
   It's important to understand that during the configuration of AAD Sync, the setup determines whether we have a Lync or Exchange deployment, and creates dedicated rules for them, as shown in the following screenshot:

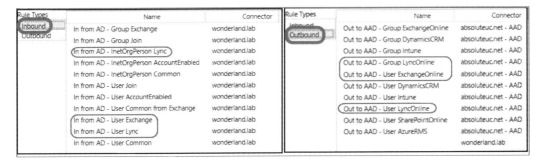

In our scenario, Lync and Exchange rules are outbound from the resource forest that hosts the previously mentioned services, and inbound to Azure, for Lync Online and Exchange Online.

3. The rules are made up from four separate pieces: **Description**, **Scoping Filter**, **Join rules**, and **Transformations**. We will see the parameters for the **In from AD - User Lync** inbound rule for the resource domain.

4. The **Description** option includes the system that will use the rule, the kind of object we apply the rule on, and the metaverse object type (this one is always a "person"). We also have to select the kind of link (**Join**, **StickyJoin**, or **Provision**). The screen for the previously mentioned rule is shown in the following screenshot:

5. The **Scoping Filter** and **Join Rules** options are empty in our rule. The former dictates when the rule has to be activated (for example, only enabled users) while the latter is used to tie an attribute to another one in the metaverse.

6. The **Transformations** option is a list of attributes that will flow (and eventually be modified) when transmitted over a connector. While the rule we are talking about has only direct transformations, with attributes mapped to the same attributes, this part of the rules can be very complex, especially when we look at an outbound rule, like the following screen, **Out to AAD - User ExchangeOnline**:

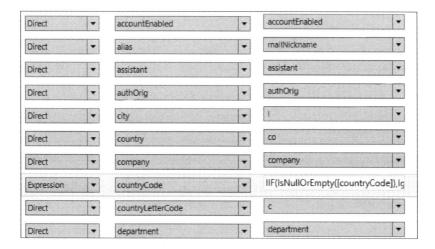

7. Now, we have to use **Synchronization Service** of AAD Sync in the start page to see the remaining parts of AAD Sync.

8. The configuration steps have created a series of **Operations** and **Connectors**, as we can see in the following screenshot:

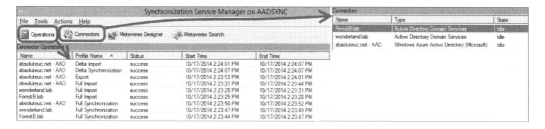

If we are already familiar with FIM, this part of AAD Sync is really similar to what we had in Identity Manager. The **Connectors** option is used to transmit data to the various systems in a transparent manner. The **Operations** option activates the transmission of information based on the rules.

9. We are able to customize the connectors. For example, we will see how to limit the organizational units used in the synchronization process. Click on **Connectors** and select the one named as the user forest (**ForestB.lab**), as shown in the following screenshot. Right-click and select **Properties**.

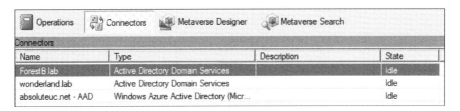

10. Select **Configure Directory Partition** and click on **Containers**. We have to insert the password required to access the **ForestB** domain and then click on **OK**. Now, we can select the **Organizational Unit** where the accounts (which we want to import in the cloud) are located, as shown in the following screenshot:

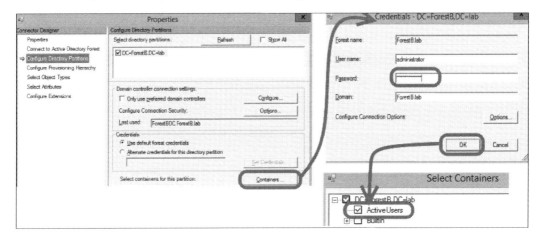

11. The steps from here on are similar to the ones that we have seen for FIM, including import, synchronization, and export to and from the metaverse.

See also

- Mike Branstein published a blog post *Using Azure Active Directory Sync to Extend your Local AD into Azure* at https://brosteins.com/2014/11/15/using-azure-active-directory-to-extend-your-local-ad-into-azure/.
- The blog of the University of Vermont contains an interesting article on *Replacing DirSync With Microsoft Azure Active Directory Sync Services* at http://blog.uvm.edu/jgm/2014/11/06/replacing-dirsync-with-microsoft-azure-active-directory-sync-services/. Their test on AAD Sync contains some important results.

8
Managing Lync 2013 Hybrid and Lync Online

In this chapter, we will cover the following recipes:

- Introducing Lync Online
- Administering with the Lync Admin Center
- Using Lync Online Remote PowerShell
- Using Lync Online cmdlets
- Introducing Lync in a hybrid scenario
- Planning and configuring a hybrid deployment
- Moving users to the cloud
- Moving users back on-premises
- Debugging Lync Online issues

Introducing Lync Online

Lync Online is part of the Office 365 offer and provides online users with the same **Instant Messaging** (**IM**), presence, and conferencing features that we would expect from an on-premises deployment of Lync Server 2013. Enterprise Voice, however, is not available on Office 365 tenants (or at least, it is available only with limitations regarding both specific Office 365 plans and geographical locations). There is no doubt that forthcoming versions of Lync and Office 365 will add what is needed to also support all the Enterprise Voice features in the cloud. Right now, the best that we are able to achieve is to move workloads, homing a part of our Lync users (the ones with no telephony requirements) in Office 365, while the remaining Lync users are homed on-premises.

Managing Lync 2013 Hybrid and Lync Online

These solutions might be interesting for several reasons, including the fact that we can avoid the costs of expanding our existing on-premises resources by moving a part of our Lync-enabled users to Office 365. The previously mentioned configuration, which involves different kinds of Lync tenants, is called a hybrid deployment of Lync, and we will see how to configure it and move our users from online to on-premises and vice versa.

 In this chapter, every time we talk about Lync Online and Office 365, we will assume that we have already configured an Office tenant.

Administering with the Lync Admin Center

Lync Online provides the **Lync Admin Center** (**LAC**), a dedicated control panel, to manage Lync settings. To open it, access the Office 365 portal and select **Service settings**, Lync, and **Manage settings in the Lync admin center**, as shown in the following screenshot:

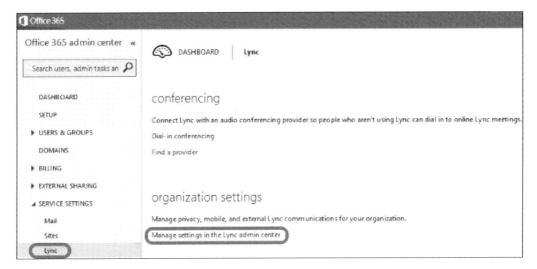

—— Chapter 8

LAC, if you compare it with the on-premises Lync Control Panel (or with the Lync Management Shell), offers few options. For example, it is not possible to create or delete users directly inside Lync. We will see some of the tasks we are able to perform in LAC, and then, we will move to the (more powerful) Remote PowerShell.

> There is an alternative path to open LAC. From the Office 365 portal, navigate to **Users & Groups** | **Active Users**. Select a user, after which you will see a **Quick Steps** area with an **Edit Lync Properties** link that will open the user-editable part of LAC.

How to do it...

1. LAC is divided into five areas: **users, organization, dial-in conferencing, meeting invitation**, and **tools**, as you can see in the following screenshot:

2. The **Users** panel will show us the configuration of the Lync Online enabled users. It is possible to modify the settings with the **Edit** option (the small pencil icon on the right):

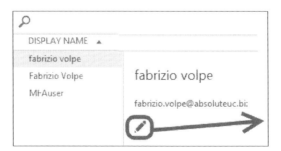

I have tried to summarize all the available options (inside the **general**, **external communications**, and **dial-in conferencing** tabs) in the following screenshot:

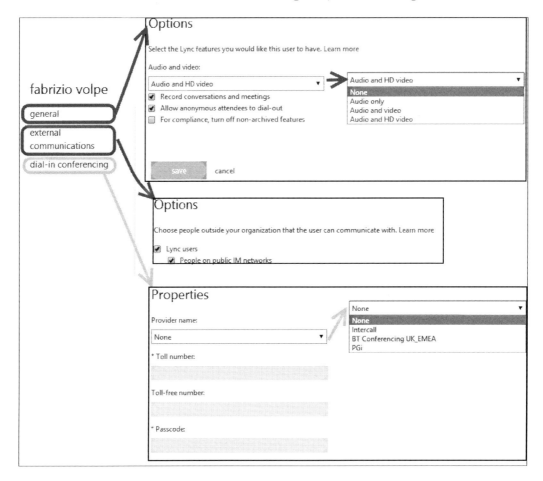

3. Some of the user's settings are worth a mention; in the **General** tab, we have the following:

    - The **Record Conversations and meetings** option enables the **Start recording** option in the Lync client
    - The **Allow anonymous attendees to dial-out** option controls whether the anonymous users that are dialing-in to a conference are required to call the conferencing service directly or are authorized for callback
    - The **For compliance, turn off non-archived features** option disables Lync features that are not recorded by In-Place Hold for Exchange

> When you place an Exchange 2013 mailbox on In-Place Hold or Litigation Hold, the Microsoft Lync 2013 content (instant messaging conversations and files shared in an online meeting) is archived in the mailbox.

In the **dial-in conferencing** tab, we have the configuration required for dial-in conferencing. The provider's drop-down menu shows a list of third parties that are able to deliver this kind of feature.

4. The **Organization** tab manages privacy for presence information, push services, and external access (the equivalent of the Lync federation on-premises). If you enable external access, we will have the option to turn on Skype federation, as we can see in the following screenshot:

public IM connectivity

☑ Turn on communication with Skype users and users of other public IM service providers.

5. The **Dial-In Conferencing** option is dedicated to the configuration of the external providers.
6. The **Meeting Invitation** option allows the user to customize the Lync Meeting invitation.
7. The **Tools** options offer a collection of troubleshooting resources.

See also

- For details about Exchange In-Place Hold, see the TechNet post *In-Place Hold and Litigation Hold* at http://technet.microsoft.com/en-us/library/ff637980(v=exchg.150).aspx.

Using Lync Online Remote PowerShell

The possibility to manage Lync using Remote PowerShell on a distant deployment has been available since Lync 2010. This feature has always required a direct connection from the management station to the Remote Lync, and a series of steps that is not always simple to set up. Lync Online supports Remote PowerShell using a dedicated (64-bit only) PowerShell module, the **Lync Online Connector**. It is used to manage online users, and it is interesting because there are many settings and automation options that are available only through PowerShell.

Managing Lync 2013 Hybrid and Lync Online

Getting ready

Lync Online Connector requires one of the following operating systems: Windows 7 (with Service Pack 1), Windows Server 2008 R2, Windows Server 2012, Windows Server 2012 R2, Windows 8, or Windows 8.1. At least PowerShell 3.0 is needed. To check it, we can use the `$PSVersionTable` variable. The result will be like the one in the following screenshot (taken on Windows 8.1, which uses PowerShell 4.0):

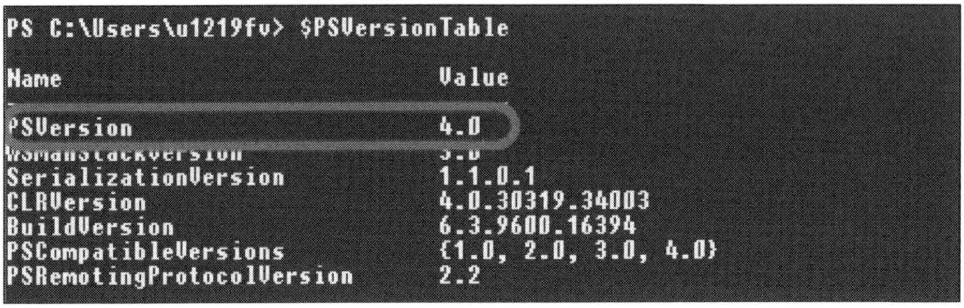

```
PS C:\Users\u1219fv> $PSVersionTable

Name                           Value
----                           -----
PSVersion                      4.0
WSManStackVersion              3.0
SerializationVersion           1.1.0.1
CLRVersion                     4.0.30319.34003
BuildVersion                   6.3.9600.16394
PSCompatibleVersions           {1.0, 2.0, 3.0, 4.0}
PSRemotingProtocolVersion      2.2
```

How to do it...

1. Download **Windows PowerShell Module for Lync Online** from the Microsoft site at http://www.microsoft.com/en-us/download/details.aspx?id=39366 and install it.

2. It is useful to store our Office 365 credentials in an object (it is possible to launch the cmdlets at step 3 anyway, and we will be required with the Office 365 administrator credentials, but using this method, we will have to insert the authentication information again every time it is required). We can use the `$credential = Get-Credential` cmdlet in a PowerShell session. We will be prompted for our username and password for Lync Online, as shown in the following screenshot:

Chapter 8

3. To use the Online Connector, open a PowerShell session and use the `New-CsOnlineSession` cmdlet. One of the ways to start a remote PowerShell session is `$session = New-CsOnlineSession -Credential $credential`.

4. Now, we need to import the session that we have created with Lync Online inside PowerShell, with the `Import-PSSession $session` cmdlet.

5. A temporary Windows PowerShell module will be created, which contains all the Lync Online cmdlets. The name of the temporary module will be similar to the one we can see in the following screenshot:

6. Now, we will have the cmdlets of the Lync Online module loaded in memory, in addition to any command that we already have available in PowerShell.

How it works...

The feature is based on a PowerShell module, the **LyncOnlineConnector**, shown in the following screenshot:

It contains only two cmdlets, the `Set-WinRMNetworkDelayMS` and `New-CsOnlineSession` cmdlets. The latter will load the required cmdlets in memory. As we have seen in the previous steps, the Online Connector adds the Lync Online PowerShell cmdlets to the ones already available. This is something we will use when talking about hybrid deployments, where we will start from the Lync Management Shell and then import the module for Lync Online. It is a good habit to verify (and close) your previous remote sessions. This can be done by selecting a specific session (using `Get-PSSession` and then pointing to a specific session with the `Remove-PSSession` statement) or closing all the existing ones with the `Get-PSSession | Remove-PSSession` cmdlet.

>
> In the previous versions of the module, Microsoft Online Services Sign-In Assistant was required. This prerequisite was removed from the latest version.

There's more...

There are some checks that we are able to perform when using the PowerShell module for Lync Online. By launching the `New-CsOnlineSession` cmdlet with the –verbose switch, we will see all the messages related to the opening of the session. The result should be similar to the one shown in the following screenshot:

Another verification comes from the `Get-Command -Module tmp_gffrkflr.ufz` command, where the module name (in this example, `tmp_gffrkflr.ufz`) is the temporary module we saw during the `Import-PSSession` step. The output of the command will show all the Lync Online cmdlets that we have loaded in memory.

The `Import-PSSession` cmdlet imports all commands except the ones that have the same name of a cmdlet that already exists in the current PowerShell session. To overwrite the existing cmdlets, we can use the `-AllowClobber` parameter.

See also

During the introduction of this section, we also discussed the possibility to administer on-premises, remote Lync Server 2013 deployment with a remote PowerShell session. John Weber has written a great post about it in his blog *Lync 2013 Remote Admin with PowerShell* at `http://tsoorad.blogspot.it/2013/10/lync-2013-remote-admin-with-powershell.html`, which is helpful if you want to use the previously mentioned feature.

Using Lync Online cmdlets

In the previous recipe, we outlined the steps required to establish a remote PowerShell session with Lync Online. We have less than 50 cmdlets, as shown in the result of the `Get-Command -Module` command in the following screenshot:

Some of them are specific for Lync Online, such as the following:

- `Get-CsAudioConferencingProvider`
- `Get-CsOnlineUser`
- `Get-CsTenant`
- `Get-CsTenantFederationConfiguration`
- `Get-CsTenantHybridConfiguration`
- `Get-CsTenantLicensingConfiguration`
- `Get-CsTenantPublicProvider`
- `New-CsEdgeAllowAllKnownDomains`
- `New-CsEdgeAllowList`
- `New-CsEdgeDomainPattern`
- `Set-CsTenantFederationConfiguration`
- `Set-CsTenantHybridConfiguration`
- `Set-CsTenantPublicProvider`
- `Update-CsTenantMeetingUrl`

All the remaining cmdlets can be used either with Lync Online or with the on-premises version of Lync Server 2013. We will see the use of some of the previously mentioned cmdlets.

How to do it...

1. The `Get-CsTenant` cmdlet will list Lync Online tenants configured for use in our organization. The output of the command includes information such as the preferred language, registrar pool, domains, and assigned plan.

2. The `Get-CsTenantHybridConfiguration` cmdlet gathers information about the hybrid configuration of Lync (we will talk about this scenario, starting with the *Introducing Lync in a hybrid scenario* section).

3. Management of the federation capability for Lync Online (the feature that enables Instant Messaging and Presence information exchange with users of other domains) is based on the allowed domain and blocked domain lists, as we can see in the **organization** and **external communications** screen of LAC, shown in the following screenshot:

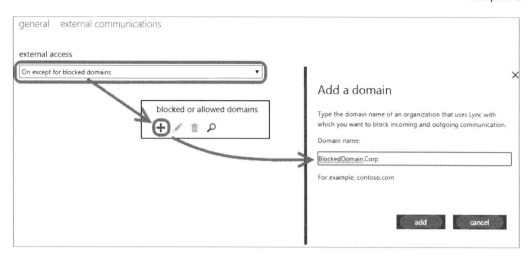

4. There are similar ways to manage federation from the Lync Online PowerShell, but it required to put together different statements as follows:

   - We can use an **accept all domains excluding the ones in the exceptions list** approach. To do this, we have put the `New-CsEdgeAllowAllKnownDomains` cmdlet inside a variable. Then, we can use the `Set-CsTenantFederationConfiguration` cmdlet to allow all the domains (except the ones in the block list) for one of our domains on a tenant. We can use the example on TechNet (http://technet.microsoft.com/en-us/library/jj994088.aspx) and integrate it with `Get-CsTenant`.

   - If we prefer, we can use a **block all domains but permit the ones in the allow list** approach. It is required to define a domain name (pattern) for every domain to allow the `New-CsEdgeDomainPattern` cmdlet, and each one of them will be saved in a variable. Then, the `New-CsEdgeAllowList` cmdlet will create a list of allowed domains from the variables. Finally, the `Set-CsTenantFederationConfiguration` cmdlet will be used. The domain we will work on will be (again) cc3b6a4e-3b6b-4ad4-90be-6faa45d05642. The example on Technet (http://technet.microsoft.com/en-us/library/jj994023.aspx) will be used:

     ```
     $x = New-CsEdgeDomainPattern -Domain "contoso.com"
     $y = New-CsEdgeDomainPattern -Domain "fabrikam.com"
     $newAllowList = New-CsEdgeAllowList -AllowedDomain $x,$y
     Set-CsTenantFederationConfiguration -Tenant " cc3b6a4e-3b6b-4ad4-90be-6faa45d05642" -AllowedDomains $newAllowList
     ```

5. The `Get-CsOnlineUser` cmdlet provides information about users enabled on Office 365. The result will show both users synced with Active Directory and users homed in the cloud. The command supports filters to limit the output; for example, the `Get-CsOnlineUser -identity` fab will gather information about the user that has alias = fab. This is an account synced from the on-premises Directory Services, so the value of the `DirSyncEnabled` parameter will be `True`.

See also

All the cmdlets of the Remote PowerShell for Lync Online are listed in the TechNet post *Lync Online cmdlets* at `http://technet.microsoft.com/en-us/library/jj994021.aspx`. This is the main source of details on the single statement.

Introducing Lync in a hybrid scenario

In a Lync hybrid deployment, we have the following:

- User accounts and related information homed in the on-premises Directory Services and replicated to Office 365.
- A part of our Lync users that consume on-premises resources and a part of them that use online (Office 365 / Lync Online) resources.
- The same (public) domain name used both online and on-premises (Lync-split DNS).
- Other Office 365 services and integration with other applications available to all our users, irrespective of where their Lync is provisioned.

One way to define Lync hybrid configuration is by using an on-premises Lync deployment federated with an Office 365 / Lync Online tenant subscription. While it is not a perfect explanation, it gives us an idea of the scenario we are talking about. Not all the features of Lync Server 2013 (especially the ones related to Enterprise Voice) are available to Lync Online users. The previously mentioned motivations, along with others (due to company policies, compliance requirements, and so on), might recommend a hybrid deployment of Lync as the best available solution. What we have to clarify now is how to make those users on different deployments talk to each other, see each other's presence status, and so on. What we will see in this section is a high-level overview of the required steps. The *Planning and configuring a hybrid deployment* recipe will provide more details about the individual steps. The list of steps here is the one required to configure a hybrid deployment, starting from Lync on-premises. In the following sections, we will also see the opposite scenario (with our initial deployment in the cloud).

How to do it...

1. It is required to have an available Office 365 tenant configuration. Our subscription has to include Lync Online.
2. We have to configure an **Active Directory Federation Services** (**AD FS**) server in our domain and make it available to the Internet using a public FQDN and an SSL certificate released from a third-party certification authority.
3. Office 365 must be enabled to synchronize with our company's Directory Services, using Active Directory Sync.
4. Our Office 365 tenant must be federated.
5. The last step is to configure Lync for a hybrid deployment.

There's more...

One of the requirements for a hybrid distribution of Lync is an on-premises deployment of Lync Server 2013 or Lync Server 2010. For Lync Server 2010, it is required to have the latest available updates installed, both on the Front Ends and on the Edge servers. It is also required to have the Lync Server 2013 administrative tools installed on a separate server.

More details about supported configuration are available on the TechNet post *Planning for Lync Server 2013 hybrid deployments* at `http://technet.microsoft.com/en-us/library/jj205403.aspx`.

> DNS SRV records for hybrid deployments, `_sipfederationtls._tcp.<domain>` and `_sip._tls.<domain>`, should point to the on-premises deployment. The `lyncdiscover.<domain>` record will point to the FQDN of the on-premises reverse proxy server. The `_sip._tls.<domain>` SRV record will resolve to the public IP of the Access Edge service of Lync on-premises.

Depending on the kind of service we are using for Lync, Exchange, and SharePoint, only a part of the features related to the integration with the additional services might be available. For example, skills search is available only if we are using Lync and SharePoint on-premises. The following TechNet post *Supported Lync Server 2013 hybrid configurations* at `http://technet.microsoft.com/en-us/library/jj945633.aspx` offers a matrix of features / service deployment combinations.

See also

Interesting information about Lync Hybrid configuration is presented in sessions available on Channel9 and coming from the Lync Conference 2014 (*Lync Online Hybrid Deep Dive* at `http://channel9.msdn.com/Events/Lync-Conference/Lync-Conference-2014/ONLI302`) and from TechEd North America 2014 (*Microsoft Lync Online Hybrid Deep Dive* at `http://channel9.msdn.com/Events/TechEd/NorthAmerica/2014/OFC-B341#fbid=`).

Planning and configuring a hybrid deployment

The planning phase for a hybrid deployment starts from a simple consideration: do we have an on-premises deployment of Lync Server? If the previously mentioned scenario is true, do we want to move users to the cloud or vice versa? Although the first situation is by far the most common, we have to also consider the case in which we have our first deployment in the cloud.

How to do it...

1. This step is all that is required for the scenario that starts from Lync Online.
    1. We have to completely deploy our Lync on-premises.
    2. Establish a remote PowerShell session with Office 365 (as we have seen in the *Using Lync Online Remote PowerShell* section).
    3. Use the shared SIP address cmdlet `Set-CsTenantFederationConfiguration -SharedSipAddressSpace $True` to enable Office 365 to use a Shared **Session Initiation Protocol** (**SIP**) address space with our on-premises deployment. To verify this, we can use the `Get-CsTenantFederationConfiguration` command. The `SharedSipAddressSpace` value should be set to `True`.
2. All the following steps are for the scenario that starts from the on-premises Lync deployment.
3. After we have subscribed with a tenant, the first step is to add the public domain we use for our Lync users to Office 365 (so that we can split it on the two deployments). To access the Office 365 portal, select **Domains**. The next step is **Specify a domain name and confirm ownership**. We will be required to type a domain name. If our domain is hosted on some specific providers (such as GoDaddy), the verification process can be automated, or we have to proceed manually. The process requires to add one DNS record (TXT or MX), like the ones shown in the following screenshot:

Record type (choose one)	Alias or Hostname	Destination or Points to Address	TTL
TXT	@ or absoluteuc.net	MS=ms31899112	1 Hour
MX	@ or absoluteuc.net	ms31899112.msv1.invalid.outlook.com	1 Hour

4. If we need to check our Office 365 and on-premises deployments before continuing with the hybrid deployment, we can use the Setup Assistant for Office 365. The tool is available inside the Office 365 portal, but we have to launch it from a domain-joined computer (the login must be performed with the domain administrative credentials). In the **Setup** menu, we have a **Quick Start** and an **Extend Your Setup** option (we have to select the second one). The process can continue installing an app or without software installation, as shown in the following screenshot:

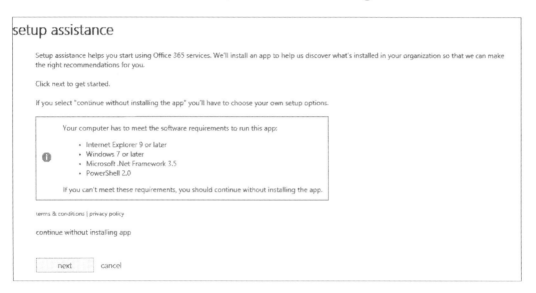

The app (which makes the assessment of the existing deployment easier) is installed by selecting **Next** in the previous screen (it requires at least Windows 7 with Service Pack 1, .NET Framework 3.5, and PowerShell 2.0).

5. Synchronization with the on-premises Active Directory is required. We can use all the information we have learned about the single sign-on test in the *Authenticating with online services using DirSync* recipe of *Chapter 2, Lync 2013 Authentication*, and test the configuration with the tools we used in the *Debugging Lync Online issues* section of this chapter.

Managing Lync 2013 Hybrid and Lync Online

6. This last step federates Lync Server 2013 with Lync Online to allow communication between our users. The first cmdlet to use is `Set-CSAccessEdgeConfiguration -AllowOutsideUsers 1 -AllowFederatedUsers 1 -UseDnsSrvRouting -EnablePartnerDiscovery 1`.

   Note that the `-EnablePartnerDiscovery` parameter is required. Setting it to 1 enables automatic discovery of federated partner domains. It is possible to set it to 0.

7. The second required cmdlet is `New-CSHostingProvider -Identity LyncOnline -ProxyFqdn "sipfed.online.lync.com" -Enabled $true -EnabledSharedAddressSpace $true -HostsOCSUsers $true -VerificationLevel UseSourceVerification -IsLocal $false -AutodiscoverUrl https://webdir.online.lync.com/Autodiscover/AutodiscoverService.svc/root`. The result of the commands is shown in the following screenshot:

```
Identity                   : LyncOnline
Name                       : LyncOnline
ProxyFqdn                  : sipfed.online.lync.com
VerificationLevel          : UseSourceVerification
Enabled                    : True
EnabledSharedAddressSpace  : True
HostsOCSUsers              : True
IsLocal                    : False
AutodiscoverUrl            : https://webdir.online.lync.com/Autodiscover/AutodiscoverService.svc/root
```

> If Lync Online is already defined, we have to use the `Set-CSHostingProvider` cmdlet, or we can remove it (`Remove-CsHostingProvider -Identity LyncOnline`) and then create it using the previously mentioned cmdlet.

There's more...

In the Lync hybrid scenario, users created in the on-premises directory are replicated to the cloud, while users generated in the cloud will not be replicated on-premises. Lync Online users are managed using the Office 365 portal, while the users on-premises are managed using the usual tools (Lync Control Panel and Lync Management Shell).

Moving users to the cloud

By moving users from Lync on-premises to the cloud, we will lose some of the parameters. The operation requires the Lync administrative tools and the PowerShell module for Lync Online to be installed on the same computer. If we install the module for Lync Online before the administrative tools for Lync 2013 Server, the `OCSCore.msi` file overwrites the `LyncOnlineConnector.ps1` file, and `New-CsOnlineSession` will require a `-TargetServer` parameter. In this situation, we have to reinstall the Lync Online module (see the following post on the Microsoft support site at http://support.microsoft.com/kb/2955287).

Getting ready

Remember that to move the user to Lync Online, they must be enabled for both Lync Server on-premises and Lync Online (so we have to assign the user a license for Lync Online by using the Office 365 portal). Users with no assigned licenses will show the error **Move-CsUser : HostedMigration fault: Error=(507), Description=(User must has an assigned license to use Lync Online**. For more details, refer to the Microsoft support site at http://support.microsoft.com/kb/2829501.

How to do it...

1. Open a new Lync Management Shell session and launch the remote session on Office 365 with the cmdlets' sequence we saw earlier. We have to add the -AllowClobber parameter so that the Lync Online module's cmdlets are able to overwrite the corresponding Lync Management Shell cmdlets:

    ```
    $credential = Get-Credential
    $session = New-CsOnlineSession -Credential $credential
    Import-PSSession $session -AllowClobber
    ```

2. Open the Lync Admin Center (as we have seen in the dedicated section) by going to **Service settings** | **Lync** | **Manage settings** in the Lync Admin Center, and copy the first part of the URL, for example, https://admin0e.online.lync.com.

3. Add the following string to the previous URL /HostedMigration/hostedmigrationservice.svc (in our example, the result will be https://admin0e.online.lync.com/ HostedMigration/hostedmigrationservice.svc).

4. The following cmdlet will move users from Lync on-premises to Lync Online. The required parameters are the identity of the Lync user and the URL that we prepared in step 2. The user identity is fabrizio.volpe@absoluteuc.biz:

    ```
    Move-CsUser -Identity fabrizio.volpe@absoluteuc.biz -Target sipfed.online.lync.com -Credential $creds -HostedMigrationOverrideUrl https://admin0e.online.lync.com/HostedMigration/hostedmigrationservice.sVc
    ```

5. Usually, we are required to insert (again) the Office 365 administrative credentials, after which we will receive a warning about the fact that we are moving our user to a different version of the service, like the one in the following screenshot:

```
PS C:\Users\administrator.WONDERLAND> Move-CsUser -Identity fabrizio.volpe@absol
uteuc.biz -Target sipfed.online.lync.com -Credential $creds -HostedMigrationOver
rideUrl https://admin0e.online.lync.com/HostedMigration/hostedmigrationservice.s
vc
WARNING: Moving a user from the current version to an earlier version (or to a
service version) can cause data loss.

Confirm
Move-CsUser
[Y] Yes  [A] Yes to All  [N] No  [L] No to All  [S] Suspend  [?] Help
(default is "Y"):
```

See the *There's more...* section of this recipe for details about user information that is migrated to Lync Online.

6. We are able to quickly verify whether the user has moved to Lync Online by using the `Get-CsUser | fl DisplayName,HostingProvider,RegistrarPool,SipAdd ress` command. On-premises **HostingProvider** is equal to **SRV:** and **RegistrarPool** is **madhatter.wonderland.lab** (the name of the internal Lync Front End). Lync Online values are **HostingProvider : sipfed.online.lync.com**, and leave **RegistrarPool** empty, as shown in the following screenshot (the user **Fabrizio** is homed on-premises, while the user **Fabrizio volpe** is homed on the cloud):

```
DisplayName      : Fabrizio
HostingProvider  : SRV:
RegistrarPool    : madhatter.wonderland.lab
SipAddress       : sip:Fab@absoluteuc.biz

DisplayName      : fabrizio volpe
HostingProvider  : sipfed.online.lync.com
RegistrarPool    :
SipAddress       : sip:fabrizio.volpe@absoluteuc.biz
```

There's more...

- If we plan to move more than one user, we have to add a selection and pipe it before the cmdlet we have already used, removing the `-identity` parameter. For example, to move all users from an **Organizational Unit** (**OU**), (for example, the LyncUsers in the **Wonderland.Lab** domain) to Lync Online, we can use `Get-CsUser -OU "OU =LyncUsers,DC=wonderland,DC=lab"| Move-CsUser -Target sipfed. online.lync.com -Credential $creds -HostedMigrationOverrideUrl https://admin0e.online.lync.com/HostedMigration/ hostedmigrationservice.sVc`. We are also able to move users based on a parameter to match using the `Get-CsUser -Filter` cmdlet.

- As we mentioned earlier, not all the user information is migrated to Lync Online. Migration contact list, groups, and access control lists are migrated, while meetings, contents, and schedules are lost. We can use the Lync Meeting Update Tool to update the meeting links (which have changed when our user's home server has changed) and automatically send updated meeting invitations to participants. There is a 64-bit version (http://www.microsoft.com/en-us/download/details.aspx?id=41656) and a 32-bit version (http://www.microsoft.com/en-us/download/details.aspx?id=41657) of the previously mentioned tool.

Moving users back on-premises

It is possible to move back users that have been moved from the on-premises Lync deployment to the cloud, and it is also possible to move on-premises users that have been defined and enabled directly in Office 365. In the latter scenario, it is important to create the user also in the on-premises domain (Directory Service).

Getting ready

1. As we have seen in other scenarios, it is required to have Active Directory synchronization and Federation Services correctly configured. Refer to the *Authenticating with online services using DirSync* recipe in *Chapter 2, Lync 2013 Authentication*.

2. It is required to federate Lync Server 2013 with Lync Online, as we have done in point 5 of the *How to do it...* section in the *Planning and configuring a hybrid deployment* recipe.

How to do it...

1. The Lync Online user must be created in the Active Directory (for example, I will define the **BornOnCloud** user that already exists in Office 365). The user must be enabled in the on-premises Lync deployment, for example, using the Lync Management Shell with the following cmdlet:

    ```
    Enable-CsUser -Identity "BornOnCloud" -SipAddress "SIP:BornOnCloud@absoluteuc.biz" -HostingProviderProxyFqdn "sipfed.online.lync.com"
    ```

2. Sync the Directory Services.

Managing Lync 2013 Hybrid and Lync Online

3. Now, we have to save our Office 365 administrative credentials in a `$cred = Get-Credential` variable and then move the user from Lync Online to the on-premises Front End using the Lync Management Shell (the `-HostedMigrationOverrideURL` parameter has the same value that we used in the previous section):

   ```
   Move-CsUser -Identity BornOnCloud@absoluteuc.biz
   -Target madhatter.wonderland.lab -Credential $cred
   -HostedMigrationOverrideURL https://admin0e.online.lync.com/
   HostedMigration/hostedmigrationservice.sVc
   ```

4. The `Get-CsUser | fl DisplayName,HostingProvider,RegistrarPool,SipAddress` cmdlet is used to verify whether the user has moved as expected.

See also

Guy Bachar has published an interesting post on his blog *Moving Users back to Lync on-premises from Lync Online* (http://guybachar.wordpress.com/2014/03/31/moving-users-back-to-lync-on-premises-from-lync-online/), where he shows how he solved some errors related to the user motion by modifying the `HostedMigrationOverrideUrl` parameter.

Debugging Lync Online issues

Getting ready

When moving from an on-premises solution to a cloud tenant, the first aspect we have to accept is that we will not have the same level of control on the deployment we had before. The tools we will list are helpful in resolving issues related to Lync Online, but the level of understanding on an issue they give to a system administrator is not the same we have with tools such as Snooper or OCSLogger. Knowing this, the more users we will move to the cloud, the more we will have to use the online instruments.

How to do it...

1. The *Set up Lync Online external communications* site on Microsoft Support (http://support.microsoft.com/common/survey.aspx?scid=sw;en;3592&showpage=1) is a guided walk-through that helps in setting up communication between our Lync Online users and external domains. The tool provides guidelines to assist in the setup of Lync Online for small to enterprise businesses. As you can see in the following screenshot, every single task is well explained:

Contact the Lync admin in the other organization

To communicate with external Lync users:

1. Contact the admin in the other organization using this email template.
2. Once they've turned on external communications, you and the other admin can add each other as Lync contacts to test things out.
3. After you and the other admin have verified that you can use Lync to communicate with each other, click **Next**.

[Next]

2. The **Remote Connectivity Analyzer** (**RCA**) (https://testconnectivity.microsoft.com/) is an outstanding tool to troubleshoot both Lync on-premises and Lync Online. The web page includes tests to analyze common errors and misconfigurations related to Microsoft services such as Exchange, Lync, and Office 365. To test different scenarios, it is necessary to use various network protocols and ports. If we are working on a firewall-protected network, using the RCA, we are also able to test services that are not directly available to us. For Lync Online, there are some tests that are especially interesting; in the Office 365 tab, the Office 365 General Tests section includes the Office 365 Lync **Domain Name Server** (**DNS**) Connectivity Test and the Office 365 Single Sign-On Test, as shown in the following screenshot:

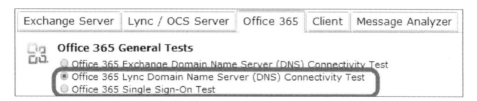

3. The Single Sign-On test is really useful in a scenario like the one we saw in the *Authenticating with online services using DirSync* recipe of *Chapter 2, Lync 2013 Authentication*. The test requires our domain username and password, both synced with the on-premises Directory Services. The steps include searching the FQDN of our AD FS server on an Internet DNS, verifying the certificate and connectivity, and then validating the token that contains the credentials. The **Client** tab offers to download the **Microsoft Connectivity Analyzer Tool** and the **Microsoft Lync Connectivity Analyzer Tool**, which we will see in the following two dedicated steps:

4. The Microsoft Connectivity Analyzer Tool makes many of the tests we see in the RCA available on our desktop. The list of prerequisites is provided in the article *Microsoft Connectivity Analyzer Tool* (`http://technet.microsoft.com/library/jj851141(v=exchg.80).aspx`), and includes Windows Vista/Windows 2008 or later versions of the operating system, .NET Framework 4.5, and an Internet browser, such as Internet Explorer, Chrome, or Firefox. For the Lync tests, a 64-bit operating system is mandatory, and the UCMA runtime 4.0 is also required (it is part of Lync Server 2013 setup, and is also available for download at `http://www.microsoft.com/en-us/download/details.aspx?id=34992`). The tools propose ways to solve different issues, and then, they run the same tests available on the RCA site. We are able to save the results in an HTML file.

5. The Microsoft Lync Connectivity Analyzer Tool is dedicated to troubleshooting the clients for mobile devices (the Lync Windows Store app and Lync apps). It tests all the required configurations, including autodiscover and webticket services. The 32-bit version is available at `http://www.microsoft.com/en-us/download/details.aspx?id=36536`, while the 64-bit version can be downloaded from `http://www.microsoft.com/en-us/download/details.aspx?id=36535`. .NET Framework 4.5 is required. The tool itself requires a few configuration parameters; we have to insert the user information that we usually add in the Lync app, and we have to use a couple of drop-down menus to describe the scenario we are testing (on-premises or Internet, and the kind of client we are going to test).

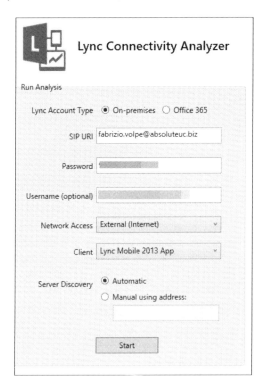

Chapter 8

The **Show** drop-down menu enables us to look not only at a summary of the test results but also at the detailed information. The detailed view includes all the information and requests sent and received during the test, with the FQDN included in the answer ticket from our services, and so on, as shown in the following screenshot:

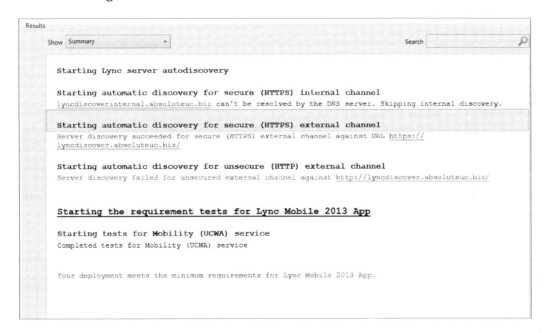

6. The *Troubleshooting Lync Online sign-in* post is a support page, available in two different versions (admins and users), and is a walk-through to help admins (or users) to troubleshoot login issues. The admin version is available at `http://support.microsoft.com/common/survey.aspx?scid=sw;en;3695&showpage=1`, while the user version is available at `http://support.microsoft.com/common/survey.aspx?scid=sw;en;3719&showpage=1`. Based on our answers to the different scenario questions, the site will propose to information or solution steps. The following screenshot is part of the resolution for the log-I issues of a company that has an enterprise subscription with a custom domain:

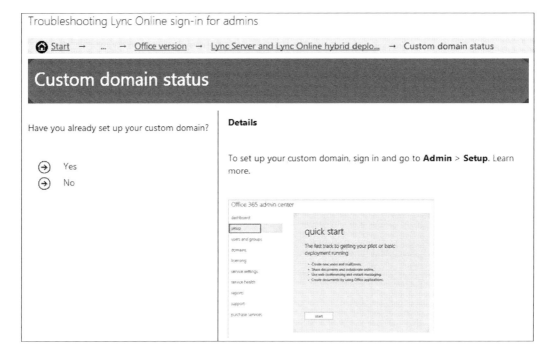

7. The Office 365 portal includes some information to help us monitor our Lync subscription. In the **Service Health** menu, navigate to **Service Health**; we have a list of all the incidents and service issues of the past days. In the **Reports** menu, we have statistics about our Office 365 consumption, including Lync. In the following screenshot, we can see the previously mentioned pages:

Chapter 8

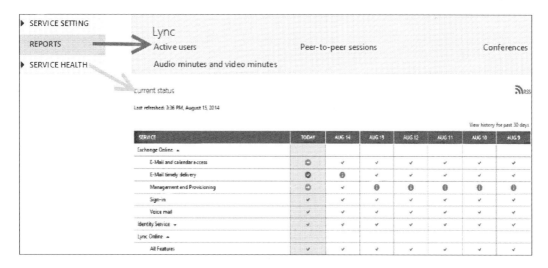

There's more...

One interesting aspect of the Microsoft Lync Connectivity Analyzer Tool that we have seen is that it enables testing for on-premises or Office 365 accounts (both testing from inside our network and from the Internet). The previously mentioned capability makes it a great tool to troubleshoot the configuration for Lync on the mobile devices that we have deployed in our internal network. This setup is usually complex, including hair-pinning and split DNS, so the diagnostic is important to quickly find misconfigured services.

See also

The *Troubleshooting Lync Sign-in Errors (Administrators)* page on Office.com at `http://office.microsoft.com/en-001/communicator-help/troubleshooting-lync-sign-in-errors-administrators-HA102759022.aspx` contains a list of messages related to sign-in errors with a suggested solution or a link to additional external resources.

9
Lync 2013 Monitoring and Reporting

In this chapter, we will cover the following topics:

- Installing Lync 2013 monitoring reports
- Selecting the right kind of report
- Call Diagnostic reports
- Media Quality Diagnostic reports
- Call Leg Media Quality report
- Lync 2013 with System Center 2012 R2 Operations Manager
- Configuring a watcher node and synthetic transactions

Introduction

Lync Monitoring is not mandatory, but it is a vital part of any deployment, to analyze and troubleshoot errors and trends (this is especially true if we are going to deploy Enterprise Voice). Using the data available through monitoring, it is possible to take design decisions and fix errors in a deployment. Monitoring Reports include reports such as user activity, conference summary, IP phone inventory, and Call Admission Control (to name a few), but the biggest benefit for Lync administrators are the diagnostic reports used to detect errors in the various functionalities of Lync. In Lync Server 2013, the monitoring functionality has been modified from the configuration point of view if we compare it with Lync Server 2010.

Lync 2013 Monitoring and Reporting

Monitoring is now a service (not a server role) co-located on the Front End Server, and it is also possible to use the Lync Back End database to store the monitoring data (*Components and topologies for monitoring in Lync Server 2013* at http://technet.microsoft.com/en-us/library/gg412952.aspx) with no dedicated SQL Server or instance required. The co-located database scenario is the easiest to implement, but it is not a recommended one for a number of reasons, including performance and reliability issues.

Installing Lync 2013 monitoring reports

We will see a quick overview of the Lync Monitoring configuration with some hints. A step-by-step guide is available both on Microsoft's TechNet (*Deploying monitoring in Lync Server 2013* at http://technet.microsoft.com/en-us/library/gg398199.aspx and on Matt Landis' site for co-located database: *Step by Step Installing Lync Server 2013 Monitoring Role Collocated on Standard Edition Front End - Part 2* at http://windowspbx.blogspot.it/2012/07/aaa-donotpost-install-lync-standard.html).

> SQL Reporting Services are not mandatory to collect Lync Monitoring data. However, they are required to make the monitoring information usable.

How to do it...

1. The first step is to install SQL Server with Reporting Services. We will install both of them on a server named marchhare.wonderland.lab.

> It is possible to install the SQL database and the Reporting Services on two different servers, but quoting TechNet (http://technet.microsoft.com/en-us/library/jj204989.aspx):
>
> *"It is recommended that you install Monitoring Reports on the same computer where the monitoring database is installed. This simplifies the process of assigning permissions for accessing the reports."*

2. As soon as the setup is completed, we are able to verify the parameter and configuration of the Reporting Services using **Reporting Services Configuration Manager**, as shown in the following screenshot (SQL Server's name is **Marchhare**, and the instance name is **Lync**):

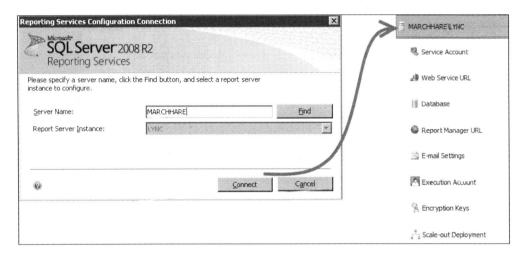

3. Now, we have to associate a Monitoring Store with a Front End Pool using **Lync Server Topology Builder**.
4. Open **Topology Builder** and download a copy of the topology (save the existing topology).
5. Expand the Lync site, Lync Server 2013, and **Standard Edition Front End Pools**.
6. Edit the Front End properties. Scroll down to **Monitoring (CDR and QoE metrics)** and check the box.

7. Next to SQL Store, click on **New**. Enter the SQL Server FQDN and the instance name and activate the flag if it is in a mirroring relationship, as we can see in the following screenshot, and then click on **OK**:

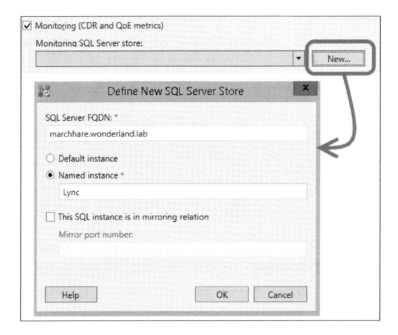

8. To save the configuration, publish the modified Lync topology.
9. The Topology Builder will define two new databases, **LcsCDR** and **QoEMetrics**, in the selected SQL instance.
10. It is recommended that we verify that the previously databases have been correctly created in the selected SQL server instance, for example, using **SQL Server Management Studio**.
11. To enable monitoring, it is required that you update the Lync Front End configuration using **Deployment Wizard** or with the `Install-CsDatabase -LocalDatabases` cmdlet.
12. We have to deploy the Monitoring Reports to the SQL Reporting Services using the **Deploy Monitor Reports** option on the right-hand side of **Lync Deployment Wizard**.

During the setup of the Monitoring Reports, it is possible to specify a Read-Only Group (enabled to access the reports with no modification rights). It is also possible to define additional roles on the reports, as outlined in the TechNet Blog post *Lync Monitoring Reports (Read-Only) Access* at `http://blogs.technet.com/b/rischwen/archive/2013/12/02/lync-monitoring-reports-read-only-access.aspx`.

13. At the end of the wizard, the URL to access the reports will be shown. In our scenario, it is `http://MARCHHARE:80/ReportServer_LYNC/Pages/ReportViewer.aspx?%2fLyncServerReports%2fReports+Home+Page`. We can copy and paste it into a browser to test the reports, as we have done in the following screenshot:

14. As a further verification that the call detail recording is enabled, we have the `Get-CsCdrConfiguration` cmdlet that will give us information about the number of day's call details and error reports to be kept in the database, as shown in the following screenshot:

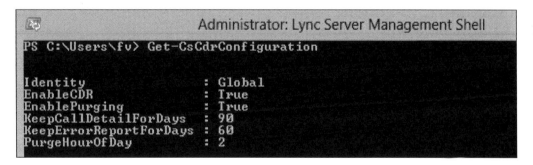

How it works...

Lync monitoring tries to detect call quality from the user's point of view (often trying to evaluate the perceived quality). After a call is concluded, the endpoint sends the **Quality of Experience** (**QoE**) data (via SIP SERVICE requests) to the Lync Front End where the **UDC agent** (component) monitors all the traffic that passes by the server. It captures the data, combines the information collected, and through an adapter, stores them in two different SQL databases (**LcsCDR** for Call Detail Records and **QoEMetrics** for Quality of Experience – QoE). All the different endpoints (clients) are able to collect the QoE data.

There's more...

The home page of Lync Server 2013 should display a link to the reporting services, as shown in the following screenshot, and it should work with no additional configuration:

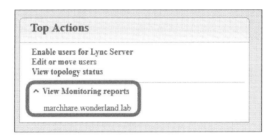

If any issue arises when launching the link, the TechNet documentation that we pointed out in the *Getting ready* section of this recipe suggests that we use the Set-CsReportingConfiguration cmdlet. Curtis Johnstone has published an interesting post (*5 Tips for Installing the Lync 2013 Monitoring Reports* at http://blog.insidelync.com/2013/05/5-tips-for-installing-the-lync-2013-monitoring-reports/) that explores this aspect of the installation with a different approach.

> If the **Lync Monitoring Reports** installation fails and SQL logs indicate **Login failed for user 'domainname\LyncServerName$'**, add the **'domainname\LyncServerName$'** account to the SQL logins and authorize it to the required databases.

See also

- Talking about Lync Monitoring, there was an interesting session by Nick Smith at the Lync Conference 2014. It was called *Getting the Most out of Lync Server Monitoring Service Data* (http://channel9.msdn.com/Events/Lync-Conference/Lync-Conference-2014/SERV302-R), and it examines the available services and helps in understanding the gathered data.

Chapter 9

- The Ignite site offers an interesting presentation (slides only), *Track 2 - Module 04 - Lync Ignite - Lync Network Assessment Methodologies*, (`http://bit.ly/1paDOmw`) on Lync monitoring.

Selecting the right kind of report

Standard reports included in Lync Server 2013 cover information on system usage, peer-to-peer sessions, conferencing sessions, and call quality. We will talk about some of them in the following paragraphs. However, in addition to the previously mentioned reports, we can also create custom reports based on the reporting data we are already gathering. Customizing reports is an important step to obtain information that fits better with our needs. We will see how to create a custom report that contains information on the time when users joined and left a conference.

How to do it...

1. Lync reports are based on **SQL Server Reporting Services** (**SSRS**). To define customized reports, we have to use built-in tools in the SQL server. We can start by opening **Reporting Services Configuration Manager** and selecting the instance that we used to install the default reports. The next step is to open the Report Manager URL in a browser window. Then, we have to select **Report Builder**, as shown in the following screenshot:

   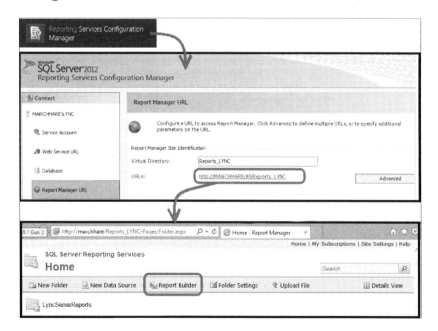

   Note that we could receive a request to download the report builder.

239

2. We must select **New Report** and then **Table or Matrix Wizard**. See the following screenshot:

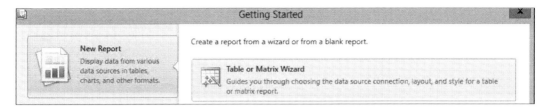

3. Now, we have to select **Create a Dataset** and click on **Next**.
4. In **Data Source Connections**, we must select **Browse** and go to the Reports_ Content folder to access the available Lync data. The selected data source screen is displayed in the following screenshot:

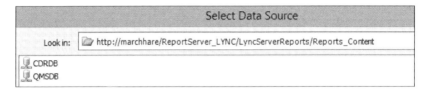

5. Select the **CDRDB** database and click on **Next**. You need to insert the credentials to access the SQL database, as we can see here:

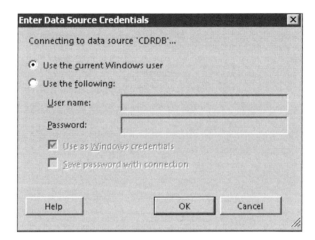

6. Select a source (**Tables**, **Views**, or **Stored Procedures**) and then select **Next**.
   We are able to select **CDRReportsMcuJoinsAndLeavesBaseView**, as shown in the next screenshot:

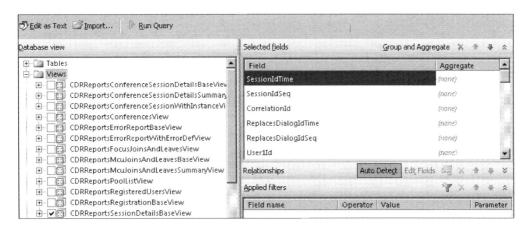

7. Click on **Next**.
8. The next phase is to arrange the fields and information, as shown in the following screenshot. We can use **UserId** in **Column groups**, **SessionIdTime** in **Row groups**, and **UserJoinTime** and **UserLeaveTime** in **Values**. For **Values**, it is required that you select a specific kind of valorization. We have to use **First**, as shown here:

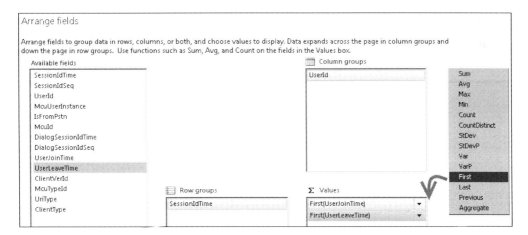

9. In the **Choose the layout screen** window, remove the **Show subtotals and grand totals** flag, as shown in the next screenshot:

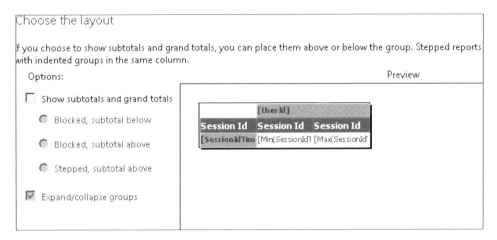

10. On the **Choose a style** screen, pick your the favorite color scheme in the example; we will use **Ocean**.
11. Now, we will have the **Reports builder** interface. Here, we can modify labels, columns, rows, and values. Every time we want to test our report, we can use the **Run** button, as shown in this screenshot:

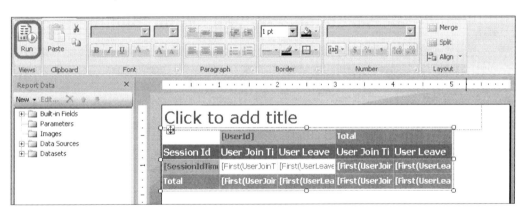

12. In our example, the result will be the one we can see in the picture. When we match **UserId** and **SessionId**, we will get a unique identification for every conference, and in the report, we will get the join and leave time for each user.

Session Id Time	User Join Time	User Leave Time	User Join Time	User Leave Time	User Join Time	User Leave Time
10/15/2014 11:03:03 AM	10/15/2014 11:03:24 AM	10/15/2014 11:05:54 AM				
10/16/2014 7:49:21 AM						
10/16/2014 8:04:42 AM					10/16/2014 8:04:46 AM	10/16/2014 8:04:57 AM
10/16/2014 8:05:45 AM					10/16/2014 8:05:45 AM	10/16/2014 8:17:02 AM
10/16/2014 8:17:07 AM					10/16/2014 8:46:50 AM	10/16/2014 8:47:01 AM
10/16/2014 12:53:41 PM					10/16/2014 12:58:52 PM	10/16/2014 1:37:59 PM
10/16/2014 1:46:52 PM					10/16/2014 1:46:53 PM	10/16/2014 1:55:41 PM

How it works...

A good starting point to write custom reports is the TechNet documentation that outlines the call detail recording (CDR) database schema (http://technet.microsoft.com/en-us/library/gg398570.aspx) and **Quality of Experience** (**QoE**) database schema in Lync Server 2013 (http://technet.microsoft.com/en-us/library/gg398687.aspx). Some of the most important views of the CDR database (suggested in the Lync Conference presentation that is already mentioned) are **Conferences**, **ConferenceSessionDetails**, **Registration**, **SessionDetails**, and **VoIPDetails**, while for the QoE database, we have **AudioStreamDetail**, **MediaLine**, **QoEReportsCallDetail**, **Session**, and **VideoStreamDetail**.

There's more...

Wesley Backelant has published a post in which he explores the use of PowerPivot as a tool to expand Lync monitoring. His work, *Extending your Lync monitoring data using PowerPivot and Power View*, is published at http://blogs.msdn.com/b/wesleyb/archive/2012/03/09/extending-your-lync-monitoring-data-using-powerpivot-and-power-view.aspx.

Iain Smith, in his blog, has dedicated an interesting post to custom reports in Lync Server 2013 (*Lync 2013 – Creating Custom Lync Reports within Monitoring* at http://northernlync.wordpress.com/2014/03/10/lync-2013-creating-custom-lync-reports-within-monitoring/).

Doug Deitterick made available a sample report dedicated to archiving (*Sample Lync Server Archiving Report Available* at http://blogs.technet.com/b/dodeitte/archive/2013/06/02/sample-lync-server-archiving-report-available.aspx).

Lync 2013 Monitoring and Reporting

Call Diagnostic Reports

The Call Diagnostic summary gives us an overview of the health status of our Lync deployment. What is remarkable about the Call Diagnostic Reports is the number of levels we are able to drill down to obtain more and more information about a single event or a single session error. While we also have other tools to log and record events, the previously mentioned monitoring report makes it easier to find the information we need.

How to do it...

1. The first level we are able to see is **Call Diagnostic Summary Report** that contains the **Peer-to-Peer Activity Diagnostic Report** and **Conference Diagnostic Report** options.
2. From the **Monitoring Server Reports** starting page, we are able to select **Call Diagnostic Summary Report**.
3. The **Diagnostic Summary** page contains the following three main fields:
    - Two fields to define a data range are shown in the following screenshot:

    - **Peer-to-peer Sessions Summary** that contains information on the success and failure in peer-to-peer (one-to-one) communication sessions is shown here:

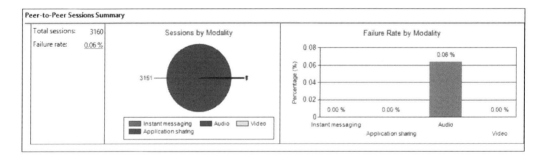

Chapter 9

- **Conferences Summary** that contains information on sessions that involve more than two users is shown in this screenshot:

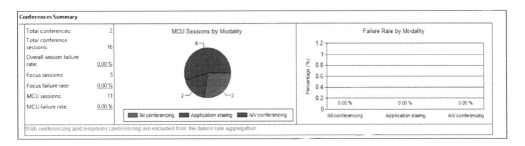

4. Looking at the pie charts in our example, we first have an overview regarding the kind of traffic we have managed in the selected period. In our examples, peer-to-peer was almost totally composed of audio calls, while the bulk part of the conference is composed of A/V Conferencing.

5. The bars on the right-hand side are the failure rates for each modality. They simply show how many sessions were not established successfully. Clicking on one of the bars will show details on the reasons behind the failure rate. In the following screenshot, we have clicked on the **Audio Failures** bar, and we are able to see a peer-to-peer activity diagnostic report that is based on day-by-day output:

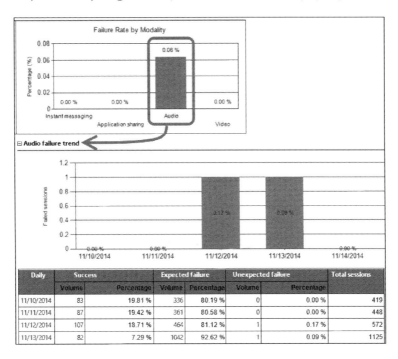

[245]

6. The way the information is divided in columns might not be intuitive. As we can see in the following screenshot, we have **Success**, **Expected Failure**, and **Unexpected Failure**:

7. Success is a call that starts and closes as expected. Expected failure includes all the sessions we were not able to establish for normal reasons (such as a call to a nonexistent phone number). Unexpected failures can arise as a result of network or device-related issues. The information is divided on a daily basis, and it helps in establishing whether the problem was due to a specific problem on a specific day or whether it was due to something in our configuration or deployment.

8. The total number of failures is also a link to **Failure Distribution Report**, as shown in the following screenshot:

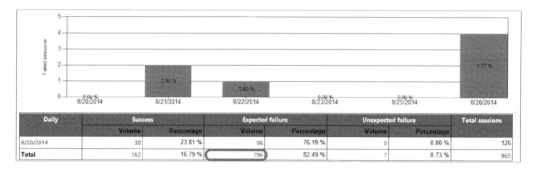

9. **Failure Distribution Report** helps us in defining trends and patterns in our errors and also in defining the specific diagnostic reason for the most frequent errors (the **Top diagnostic reasons** table is shown in the following screenshot). The **Top pools** table shows how failures are distributed among our Lync Front End servers.

Rank	Top diagnostic reasons	Sessions
1	22 - Call failed to establish due to a media connectivity failure w...	21
2	51007 - Callee media connectivity diagnosis info	18
3	7037 - Media stack diagnostics info	12
4	10040 - Unexpected call termination from gateway side, ITU-T Q.850 ...	10
5	10503 - Gateway responded with 503 Service Unavailable	8
6	10001 - Gateway did not respond in a timely manner (timeout)	6
7	24041 - No message received handler	5
8	10006 - Proxy side Media negotiation failed	1
9	1045 - Local edge server pool is out of service	1
10	21009 - Media stack diagnostics info	1

Chapter 9

On the same page, we also have the **Top Sources**, **Top Components**, **Top from users**, **Top to users**, and **Top from user agents** tables.

10. Clicking on the number of events to the right of each element, we can see that the number of event is a link to another level of reporting:

11. Now, we can examine every single event related, for example, to a calling user number (see the following screenshot). Note that there is a **Session Detail** link that allows us to go to a further level of detail:

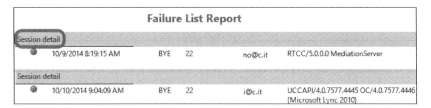

12. The **Peer-To-Peer Session Detail Report** option is an in-depth look at the session detail error. In the following screenshot, we also have a demonstration of the "highlighting" feature that points out errors using yellow and red colors. Hovering the mouse over the errors will bring up information on this kind of message.

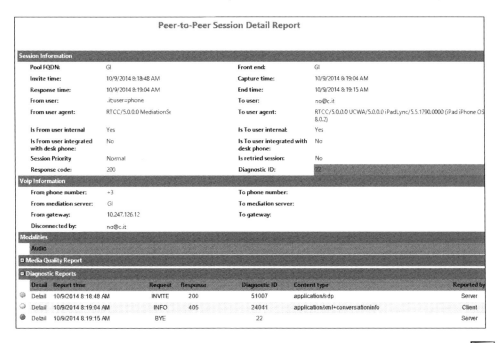

Lync 2013 Monitoring and Reporting

13. In the **Diagnostic Reports** tab, there can be additional details linked in the same manner, as shown in the following screenshot:

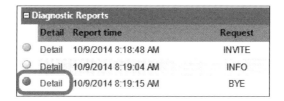

There's more...

The *explanation while hovering on information* we have seen in the **Peer-To-Peer Session Detail Report** feature is available in all the Lync reports, as we can see in this screenshot:

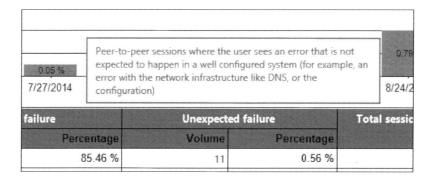

Media Quality Diagnostic Reports

Lync reports enable more than one approach to the assessment of errors and trends in call quality. The starting page of the monitoring reports offers two different starting points for this kind of activity: **Call Diagnostic Reports (per user)** and **Media Quality Diagnostic Reports** (as shown in the following screenshot). We will look at both of them.

Chapter 9

How to do it...

1. The **User Activity Report** option contains a full list of all the users' activities in the defined range of time. We have some filters available. In the following screenshot, I have pointed out the drop-down menus: the one related to **Activity Type** (the available options are **Peer-to-peer**, **Conference**, or **Both**) and the one related to the **Session Category** (the available options are **Success**, **Expected Failure**, **Unexpected Failure**, or **All**). The view is the one that shows unexpected failures in all the category types.

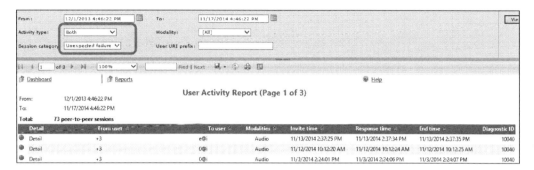

2. User activity reports give access to every single activity performed by a specific user. The **Detail** link offers access to a detailed report screen that includes **Media Quality Report**; this includes information on the call and details on the audio and video quality.

3. **Media Quality Diagnostic Reports** include the following reports:
    - Media Quality Summary Report
    - Media Quality Comparison Report
    - Server Performance Report
    - Call Leg Media Quality Report
    - Location Report
    - Device Report

4. The **Media Quality Summary Report** option includes many quality-related counters such as **MOS** (Mean Opinion Score), which is a methodology for subjective testing (because the quality of speech is based on a human perception). Other parameters include **Jitter** (a measure of distorted or lost audio, typically caused by network congestion) and **Healer concealed ratio** (usually caused due to packet loss or jitter). In the following screenshot, we have a high number of lost packets that cause an increase in **Healer concealed ratio**:

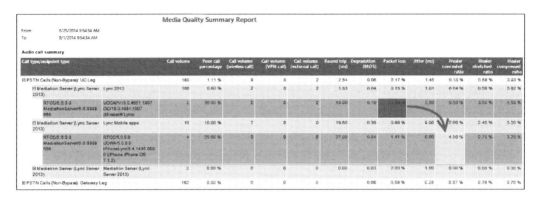

5. The **Percentage of Packet Loss** option is a link to **Media Quality Metrics Distribution Report** and after selecting **Call List**, we have a further step, the **Call List Report**. Inside the **Call List Report** option, we are able to see that in our example, there was a single call with a high number of lost packets, as shown in the following screenshot:

6. The **Media Quality Comparison Report** option is a view of daily calls from multiple points of view (such as **Call Volume**, **MOS**, and **Poor call percentage**).

7. We will talk about **Call Leg Media Quality Report** later in this chapter. The last report that we will talk about is **Device Report**; it reports the quality of the calls related to the different devices that were used, as illustrated here:

	Device Report		
From:	8/25/2014 10:26:51 AM		
To:	9/1/2014 10:26:51 AM		

Capture device	Render device	Call volume	Poor call percentage
Summary			
Microfono interno (Conexant 20672 SmartAudio HD)	Altoparlanti (Conexant 20672 SmartAudio HD)		0.00 %
Cuffia auricolare con microfono (Plantronics BT300M)	Auricolare e microtelefono (Plantronics BT300M)		0.00 %
Cuffia auricolare con microfono (Plantronics C325-M)	Auricolare e microtelefono (Plantronics C325-M)		0.00 %
iOS Microphone	iOS Speaker		0.00 %
Transmit (2- Plantronics D100-M)	Receive (2- Plantronics D100-M)		0.00 %
Transmit (Plantronics D100-M)	Receive (Plantronics D100-M)		0.00 %

There's more...

Metrics and details about the User Activity Report are available on the TechNet post, *User Activity Report in Lync Server 2013*, at `http://technet.microsoft.com/en-us/library/gg558638.aspx`. More information on MOS is available in the TechNet post, *Mean Opinion Score and Metrics*, at `http://technet.microsoft.com/en-us/library/bb894481(v=office.12).aspx`.

See also

- The *Lync 2013 Rollout and Adoption Success Kit* post at `http://www.microsoft.com/en-us/download/details.aspx?id=37031` includes the Lync Pilot Deployment Health Analysis Tool. It is an instrument that helps in identifying Lync errors and evaluating call quality. Michael LaMontagne has published an interesting post on this topic, *Lync Pilot Deployment Health Analysis Tool*, at `http://d1it.wordpress.com/2013/10/24/lync-pilot-deployment-health-analysis-tool/`.
- Curtis Johnstone has published an interesting post (focused mainly on the audio-quality side of the reports) on Lync monitoring, titled *A Primer on Lync Audio Quality Metrics* at `http://blog.insidelync.com/2012/06/a-primer-on-lync-audio-quality-metrics/`.

Call Leg Media Quality Report

A call leg is a logical connection between two telephony devices, regardless of their type. **Call Leg Media Quality Report** is dedicated to the quality of the communication channel between the Lync Mediation Server role and an external PSTN gateway (or an external **Audio/Video Multi-Point Conference Unit** (**AVMCU**)).

Quoting the Microsoft Support post at http://support.microsoft.com/kb/2798143:

> "An update is available that enables you to use a Call Leg Media Quality Report in a Lync Server 2013 environment.... The report also enables you to identify and resolve issues of audio quality that are caused by media server connection issues."

How to do it...

1. When we open the **Call Leg Media Quality Report** page, we have a list of connections between our Mediation server and the other media servers. In the following screenshot, we have two SIP trunks that connect Lync to two **Cisco Unified Communications Manager** (**CUCM**):

Endpoint 1	Endpoint 2	Call volume	Poor call percentage	Round trip (ms)	Degradation (MOS)	Packet loss	Jitter (ms)
02	trunk=Tr13	151	0.66 %		0.09	0.86 %	0
01	trunk=Tr12	133	0.00 %		0.10	0.86 %	0

2. On the right-hand side of the **Call Leg Media Quality Report** page, there is a **Trend** link that opens the **Call Leg Media Quality Trend Report** page; this page will show daily overviews about Call Volume, MOS, Jitter, and other parameters, as shown in this screenshot:

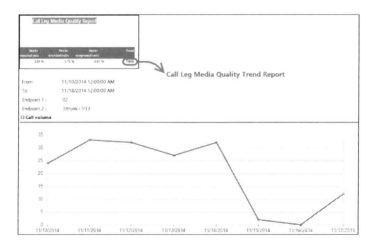

3. Opening the **Call Volume** links, we will be routed to the **Call List Report** page. Any call with quality issues will be pointed out with a yellow or red color, like the ones we are able to see here:

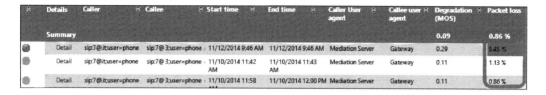

4. When we select **Detail**, a **Call Detail Report** page will show information on the call, the server and gateways involved, used bandwidth, and so on.

There's more...

Usually, customers need to use reports to recharge call costs. Call accounting is an important aspect for many companies, especially if there is a need to recharge the costs to the various departments/offices. There are a lot of third-party solutions, but they are often costly and add features that not every business is interested in. In this section, we will simply point out a couple of free solutions coming from the Microsoft Lync community. Details are available on the sites of the authors.

Andrew Morpeth has published a Lync Call Report that processes data from the Lync Monitoring Server database (LcsCDR database) and allows to import cost rates and association of costs to the different departments. The report is titled *Call Accounting* and is available at http://www.lync.geek.nz/p/call-accounting.html.

Frederik Lefevre has published *Lync CDR Report* at http://www.ucprofessional.com/2012/01/lync-cdr-report.html. It is not a preconfigured report, but an explanation on how to build our own CDR report. We have to consider it together with the *Lync 2013 CDR Report – Recharge Call Costs* post from Chris Hayward at http://weakestlync.com/2013/07/11/lync-2013-cdr-report-recharge-call-costs/. This post adapts the information from Frederik Lefevre to Lync Server 2013.

Lync 2013 with System Center 2012 R2 Operations Manager

System Center 2012 R2 Operations Manager (**SCOM**) is used to monitor the health and performance in a network environment, especially around Microsoft products. From the point of view of a Lync administrator, SCOM automates many operational and administrative tasks that involve monitoring and management. It gathers data and logs using the agents installed on the network servers and composes them in a series of reports and warnings. When we select a series of thresholds, we will be able to start automatic alerts and (also) automatic remediation actions (often integrating SCOM with other products in the System Center suite). SCOM is also able to watch the status of Lync external dependencies and integrate with the Lync Monitoring service we talked about in previous sections. The default installation of Operations Manager is able to monitor only a small number of events and requires the installation of management packs (software that contain monitoring tools and point out to SCOM what objects have to be watched and how to monitor them).

Getting ready

There are two management packs for Lync Server 2013: **Microsoft Lync Server 2013 Management Pack** and **Microsoft Lync Server 2013 Remote Watcher Management Pack**. We will focus on the first one, while the Remote Watcher Management Pack (and **Watcher Node**) will be the topic of the next section. We will talk about SCOM and Lync 2013 integration, so details about the SCOM setup and configuration are not included in the text. However, there are a lot of resources available on this topic, for example *The Complete Home Lync Lab: Part 5 Installing System Center Operation Manager 2012 R2* post by Andrew Price at http://lyncme.co.uk/microsoft-lync-server-2013/the-complete-home-lync-lab-part-5-installing-system-center-operation-manager-2012-r2/.

How to do it...

1. Add the Lync Servers to the computers managed with SCOM (if it is not already monitored) by launching **Operation Manager**, **Device Management**, and **Configure Computers and Devices to Manage**. Select **Windows Computer**, as shown in this screenshot:

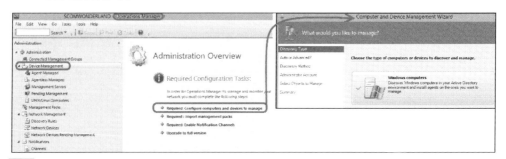

2. Select the way to scan for computers (**Automatic** or **Advanced**) and then click on **Next**.
3. Select the administrative account that will scan the newly-found computers and select **Discover**.
4. Select the computers you want to manage, including the ones with services required for Lync, such as the Office Web Apps server, and select **Next**, as shown here:

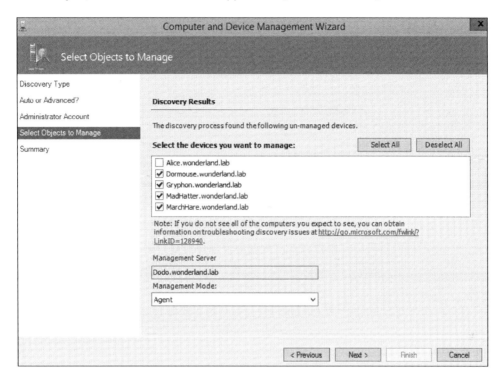

 To monitor Lync Edge servers, we have to use the steps required to configure the Agent on a workgroup server; these steps are based on certificates. The procedure creates a really secure connection between the SCOM server and the machine in the DMZ, based on encryption and requiring a single TCP port (5273) to work. For more information, refer to the TechNet blog post, *Monitoring non-domain members with OM 2012*, at `http://blogs.technet.com/b/stefan_stranger/archive/2012/04/17/monitoring-non-domain-members-with-om-2012.aspx`.

5. Select the **Agent installation** directory and the **Agent Credentials** and then click on **Finish**.

Lync 2013 Monitoring and Reporting

6. Wait for **Agent Management Task Status** to flag the agent installation as complete.
7. Go to **Authoring** (or to **Administration**) in **Operation Manager** and select **Import Management Packs**:

8. On the Import Management Packs screen, select **Add from Catalogue**. Select the Lync-related management packs (Microsoft Lync Server 2013 Management Pack and Microsoft Lync Server 2013 Remote Watcher Management Pack) and click on Add (as shown in the next screenshot). This is a good moment to add other management packs we could be missing (for example, to watch for Lync-related services such as IIS or File Services if we have our Lync share on a DFS).

9. On the next screen, select **Install**. Note that if we are using System Center 2012 Operations Manager, we have to download the management packs and install them manually, as explained in the TechNet post, *Importing the Lync Server 2013 management packs*, at http://technet.microsoft.com/en-us/library/jj205052.aspx.
10. After a while, the Lync Servers will be shown as **Healthy** in the **Monitoring | Discovered Inventory** view. We will have two additional views by navigating to **Operations Manager Console | Monitoring**. The first one is **Microsoft Lync Server 2013 Health**, and the second one is **Microsoft Lync Server 2013 Remote Watcher Health** (each one is related to one of the management packs we installed in the previous steps). However, the views will be empty, because an additional step is required, as shown in the following screenshot:

Chapter 9

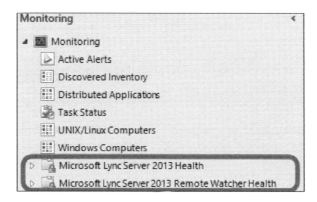

11. Go to **Administration** | **Agent Managed**, right-click on your Lync Servers, and select **Properties**.

12. Go to the **Security** tab and select **Allow this agent to act as a proxy and discover managed objects on other computers**, as shown here:

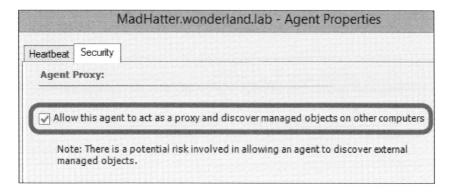

 The previously mentioned actions have to be repeated on all the servers we are going to monitor. It can take up to 4 hours before the new configuration is received. Restarting the health services will speed up this process (`net stop healthservice && net start healthservice`).

257

Lync 2013 Monitoring and Reporting

13. As soon as the SCOM agents update the information, we will be able to check many aspects of our Lync deployment health. Have a look at the following examples:

    - The **Microsoft Lync Server 2013 Health** option includes some "first level" views, such as the **Deployment Diagram** option that will show an outline of our Lync deployment with error messages:

    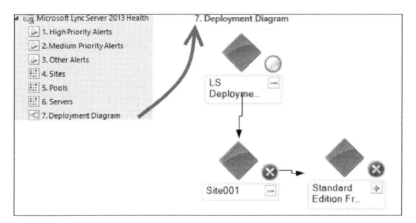

    - There is an additional level of views grouped by topic. The **Components** view includes a series of checks dedicated to the different base services of Lync. In the following screenshot, we are able to see the status of **Core Service**:

    - The **Servers** view separates the information based on the Lync Server role:

    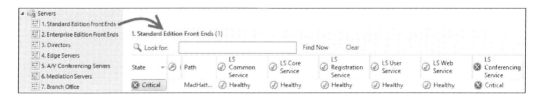

❑ The **Call Reliability and Media Quality** view uses metrics measured by call detail recording and QoE:

There's more...

The Lync Edge Server is not part of the domain. To monitor it, it's required that you use a process like the one outlined in the *Monitoring non-domain members with OM 2012* post at http://blogs.technet.com/b/stefan_stranger/archive/2012/04/17/monitoring-non-domain-members-with-om-2012.aspx.

The Lync Server 2013 Management Pack Guide (that is included in the Lync Server 2013 Management Pack at http://www.microsoft.com/en-us/download/details.aspx?id=35842) explains additional details about the management packs, including the requirement to configure each server to act as a proxy.

See also

▶ Stale Hansen, during the **Nordic Infrastructure Conference** (**NIC**) Conference 2013, presented an interesting session regarding SCOM (in its 2012 release) and Lync active monitoring. This session was called *Lync Server 2013: Best Practice Administration With System Center Operations Manager 2012*. It is available at http://vimeo.com/58166528.

▶ A post is dedicated to active monitoring on the TechNet (formerly, NextHop) blog, *Notes from the Field: Lync Server 2013 and Active Monitoring with System Center Operations Manager*, at http://blogs.technet.com/b/nexthop/archive/2013/05/17/notes-from-the-field-lync-server-2013-and-active-monitoring-with-system-center-operations-manager.aspx.

Configuring a watcher node and synthetic transactions

Synthetic transactions are PowerShell cmdlets built into Lync Server that are used to simulate activities performed by servers or users in our Lync deployment. They are a way to test a Lync service in an accurate way. Although it is possible to use synthetic transactions manually, the Lync management pack for SCOM, **Microsoft Lync Server 2013 Remote Watcher Management Pack** (that we introduced in the previous section), enables the use of synthetic transactions for active and automated monitoring of our Lync infrastructure. The previously mentioned feature requires a server (called watcher node) that will actually periodically run the synthetic transactions. A watcher node is required for every Lync site that we want to monitor with the previously mentioned feature.

How to do it...

1. A watcher node server has some prerequirements: .NET Framework 4.5., **Windows Identity Foundation** (**WIF**), and Windows PowerShell 3.0. In Windows Server 2012 R2, only WIF is not enabled by default. We can enable it from PowerShell using the `add-WindowsFeature windows-identity-foundation` cmdlet.

2. We have to mount the Lync Server 2013 installation media (for example, with the `D:` letter) and install the core files and the RTCLocal database using the `D:\setup\amd64\Setup.exe /BootstrapLocalMgmt` cmdlet.

3. Update the Lync components we have just installed according to your servers' patch level.

4. Install the Local Configuration Store from **Lync Server 2013 – Deployment Wizard**, as shown in the following screenshot:

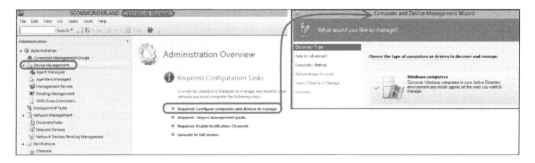

5. Install the SCOM agent and enable it to act as a proxy, as we saw in the previous section.

6. Assign a default certificate to each watcher node, using **Lync Server Deployment Wizard**. From **Lync Server 2013 – Deployment Wizard**, go to **Install or Update Lync Server System | Step 3 Request, Install or Assign Certificates**. In the following series of screenshots, we can see some of the usual steps to request a certificate to an internal certification authority:

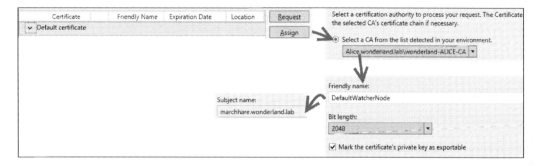

7. We have to assign the default certificate generated in the previous step.

8. Create a Trusted Server Application Pool from the Lync Management Shell (note that **marchhare.wonderland.lab** is the watcher node and **madhatter.wonderland.lab** is the Lync Front End):

   ```
   New-CsTrustedApplicationPool -Identity marchhare.wonderland.
   lab -Registrar madhatter.wonderland.lab -ThrottleAsServer
   $True -TreatAsAuthenticated $True -OutboundOnly $False -
   RequiresReplication $True -ComputerFqdn marchhare.wonderland.lab
   -Site Site001
   ```

9. Create a Trusted Application:

   ```
   New-CsTrustedApplication -ApplicationId STWatcherNode -
   TrustedApplicationPoolFqdn marchhare.wonderland.lab -Port 5061
   ```

10. Enable the changes we made outside the **Topology Builder** using the `Enable-CsTopology` cmdlet and restart **Health service** (`net stop healthservice && net start healthservice`).

11. Download the watcher node file, `WatcherNode.msi`, included in the Lync Server 2013 Management Pack (`http://www.microsoft.com/en-us/download/details.aspx?id=35842`) and copy it to a folder (for example, `c:\wn`). Launch the installation with the following command:

    ```
    C:\wn\WatcherNode.msi Authentication=TrustedServer
    ```

12. Configure the Lync watcher node. It is required that you have some test users with the right voice policies assigned (in our example, Watcher1@absoluteuc.biz, Watcher2@absoluteuc.biz, and Watcher3@absoluteuc.biz):

```
New-CsWatcherNodeConfiguration -TargetFqdn "LYNCSITEPOOL.CORP.
COM" -PortNumber 5061 -TestUsers @{Add= "sip: Watcher1@absoluteuc.
biz","sip: Watcher2@absoluteuc.biz", "sip: Watcher3@absoluteuc.
biz"}
```

There's more...

Corrado Mollica has published (in an Italian blog) an interesting script, `WatcherNode.ps1`, that automatizes a part of the configuration process. It is available at http://blogs.pulsarit.net/2014/06/watchernode-ps1/.

10 Managing Lync 2013 Backup and Restore

In this chapter, we will cover the following backup and restore topics:

- Topology information
- Configuration information
- User database
- Persistent Chat database
- The Location Information Service (LIS) database
- The Response Group Service (RGS) configuration
- Certificates
- Backend databases
- Voice dial plans, policies, and settings
- File services
- Don't forget the infrastructure – the greater recovery plan

Introduction

The Lync Server 2013 infrastructure is an important part of any organization. Just like any other infrastructure or system, it is important to have a sound backup routine and a well thought through recovery plan.

Managing Lync 2013 Backup and Restore

Even with a fully redundant Lync architecture built to Microsoft Best Practice for High Availability and Failover (http://technet.microsoft.com/en-us/library/jj204703.aspx), we could still be faced with failed replication or database corruption. In departments with many resources, there is also a risk that the configuration or topology could be overwritten unintentionally, or we might just be really unlucky and lose everything.

However unlikely it is, it is critical that there is a proper disaster recovery plan available, including backup and restore. This is what this chapter is all about.

There might be more than one way to create a backup and more than one way to restore. I will focus on the best practice, if there is one, in my walk-through and will talk about other viable options in the *There's more* sections as we go along.

The *only* way to be certain that our backups are OK and usable for disaster recovery is to actually try to restore the services in a lab environment from scratch. Practicing this will also prepare us for the real deal when a disaster strikes.

Some of the restore jobs will require a restart of servers or services but are not included in the *How to do it...* sections. The rule of thumb is that to ensure stability and continued functionality, it is recommended that you perform a restart of the services involved.

A prerequisite to installing Lync 2013 is to install PowerShell v3. All modules should load automatically. If this fails for some reason, run **Import-module Lync** to manually load the module.

Topology information

The topology is the heart of the Lync installation. In the topology, there is information about sites, pools, servers, file stores, different databases, domains, services, and roles.

Without this piece of information, we will not be able to restore anything, and we will have to recreate the entire installation from scratch.

Creating a backup of your topology is an easy and quick task.

Getting ready

There are two ways to back up or restore your topology. The supported way according to Microsoft is to do it through the **Topology Builder**.

The Topology builder is installed together with the **Lync administrative tools** and must be run by a user with administrative privileges on the system, typically a member of the **CSAdministrator** group.

How to do it...

The following are five easy steps to back up the Topology:

1. Launch **Topology Builder**.
2. Once it launches, make sure that **Download Topology from existing deployment** is selected as follows:

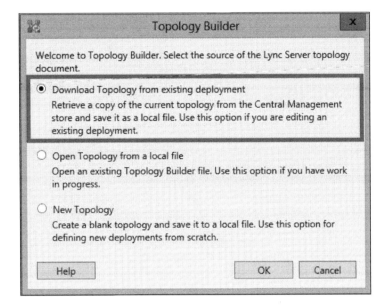

3. Click on **OK**, and select the path and filename when you save the file.

The file will be a `.tbxml` file type, and we must use Topology Builder to restore the installation.

1. Close down Topology Builder.
2. Move the file to a safe location.

Managing Lync 2013 Backup and Restore

> When editing a topology by adding, removing, or changing objects, we can make a copy of this file before we start editing the topology. This way, we will have a short way to restore the system (or compare before and after changes) if any of the changes broke something.
>
> If possible, it could be wise to map a drive to a central file server and adjust the commands in this chapter to store files in a fileshare covered by a regular file backup plan.

The following are the steps to restore the topology:

1. Open Topology Builder.
2. Once it launches, select the **Open Topology from a local file** option and then click on **OK**, as shown in the following screenshot:

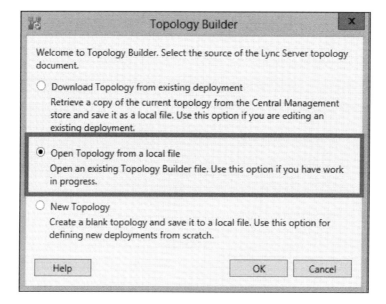

3. When we click on **OK**, we have to then browse to the file location and select the file.
4. Go through the entire deployment and verify everything before thinking of publishing it further.

Chapter 10

5. When we're satisfied with how everything looks, we can go ahead and publish the topology, as shown in the following screenshot. Before we do so, all SQL server databases and instances must be configured and ready to receive data. The same goes for the file server (we will get a warning message about this when we publish from the file; read it and make sure that all the bullet points are checked).

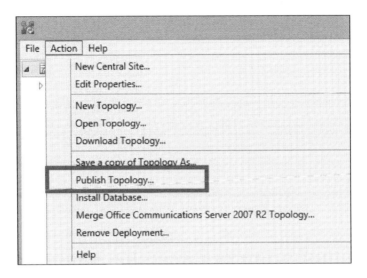

6. If all is right, we should now have a successful restore of the Lync Topology.
7. Wait for the Lync replication to complete.

> To verify the replication status, we can run `Get-CsManagementStoreReplicationStatus` on the server.
>
> For a quick view of the failed replication, we can run the following command:
>
> `Get-CsManagementStoreReplicationStatus | ft -property uptodate,replicafqdn -auto`

267

Managing Lync 2013 Backup and Restore

8. Run bootstrapper on any server where changes require such an action. Open the **Click here to open to-do list** option on the result page of **Publish topology** wizard, as shown in the following screenshot:

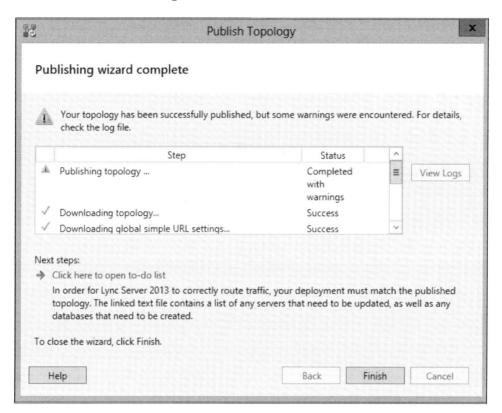

 Bootstrapper is the process of installing or removing all the (un)configured components of the Lync Server. It is a simple DOS command. Simply run `Bootstrapper.exe` from the path it is installed in. The default installation path is `C:\Program Files\Microsoft Lync Server 2013\Deployment\`.

Chapter 10

The file should look something like the following screenshot:

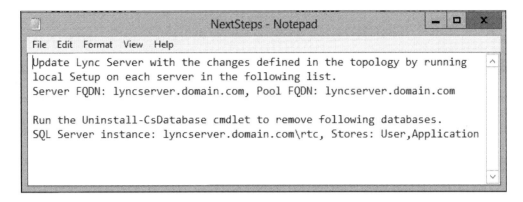

9. Depending on what we are restoring, run the suggested steps under **Install or Update Lync Server System** in **Lync Server 2013 – Deployment Wizard**.

 If we have made changes to the topology and are trying to revert a change, we would get a warning about this, and we should be absolutely sure that we are doing the right thing before we go on.

There's more...

The preceding method is just one way to create a backup and restore it. It is fairly simple and quite safe. However, it is a manual process, and in my experience, it tends to be forgotten every now and then.

If we want to automate this procedure (taking a backup), we can easily do so with the `Get-CsTopology` command in a PowerShell script and have this script run as a scheduled task. The trick about this command is that it usually just returns the information in the PowerShell window, and this is no good for us. We need to transform the information into an XML file like the following one:

`(Get-CsTopology -AsXml).ToString() > C:\Backup\Topology-backup.xml`

As this file cannot be imported into the Topology Builder due to changes made in Lync 2013, we will also have to rely on PowerShell for the restore procedure:

`Publish-CsTopology -FileName "C:\Backup\Topology-backup.xml"`

Remember that all SQL databases, filestores, and machine accounts must be deployed and operative before we do this; otherwise, the publication will fail.

269

To finalize the restoration, we will have to run one more command:

`Enable-CsTopology`

See also

- Read about `Get-cstopology` at http://technet.microsoft.com/en-us/library/gg412824.aspx
- Read about `Publish-cstopology` at http://technet.microsoft.com/en-us/library/gg398953.aspx
- Read about `Enable-cstopology` at http://technet.microsoft.com/en-us/library/gg398398.aspx

Configuration information

The next topic is the second most important piece of the puzzle in a Lync installation. It is a backup of the entire configuration store from the Central Management Store. Within this file lies the entire topology, list of servers, policies, settings, and other important information.

The instructions for backup and restore do not differ much from the instructions of the Edge Server topology deployment or any other non-domain-joined server.

Getting ready

PowerShell is the chosen tool for this task, and someone who is a member of the **RTCUniversalServerAdmins** group must run the command.

The Lync Core software must be installed to complete these tasks.

How to do it...

The following are the three steps to back up the entire configuration:

1. Start PowerShell in the elevated administrative mode.
2. Type `Export-CsConfiguration -FileName "C:\Backup\LyncConfigBackUp.zip"`.
3. Save the file in a secure location.

It could not get any easier than this.

The following are the three steps to restore the entire configuration:

1. Start PowerShell in the elevated administrative mode.
2. Type `Import-CsConfiguration -FileName "C:\Backup\LyncConfigBackUp.zip"`.
3. Run the deployment wizard on the Lync Server that we want to restore first.

How it works...

The Central Management Store has now been updated with the configuration we earlier stored in the backup file. Depending on the status of the rest of the installations, we might have to wait for a Lync replication before we start running bootstrapper on all servers. See the *Restore topology information* section for tips on checking replication with the use of **Get-CsManagementStoreReplicationStatus**.

In the event of a total disaster, this might be the first server we restore, and there will be no servers to replicate with at the moment. Any new server from which we run the deployment wizard will fetch its configuration directly from the newly-restored information.

After running the bootstraper, the next step will be to add certificates to the services before starting services.

There's more...

The information stored in the export of the configuration can be used for purposes other than system restoration. This process is (as I mentioned) the method to implement a Lync Server that is not a member of the domain and cannot read the Active Directory to find the location of the Central Management Database. The Edge server is one such server, but it can be another third-party-trusted application server as well.

On a server that is not part of the AD domain, use `-Localstore` in the import command, as follows:

`Import-CsConfiguration -FileName "C:\Backup\LyncConfigBackUp.zip" -Localstore`

The next move is to run the bootstrapper to deploy services intended for this computer.

After bootstrapper has completed, it is time to assign the certificates before starting the services.

Managing Lync 2013 Backup and Restore

The task of importing the file to the local store on a server can also be completed in the **Lync Server 2013 Deployment Wizard** GUI as follows:

1. Start the **Lync Server 2013 Deployment Wizard** window.
2. Select **Install or Update Lync Server System**, as shown in the following screenshot:

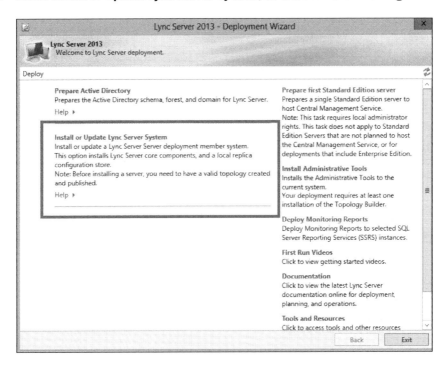

3. Then, select **Run** under the **Step 1: Install Local Configuration Store** option.
4. If the server cannot locate the Central Management Store, it will prompt you for the input of a configuration file. Browse and select the desired file, as shown in the following screenshot. This method will use the `-localstore` switch by default, and it cannot be used to recreate the entire Central Management Store as described earlier.

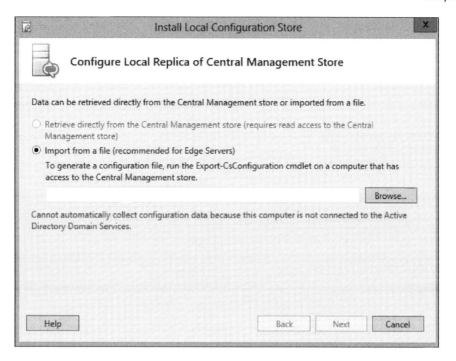

5. Follow the rest of the steps in the deployment wizard, and the server will be ready for production.

 This last method can be used to restore (or rather prepare) a server in a lab environment before moving it into production.

See also

- The *Export-CsConfiguration* post at `http://technet.microsoft.com/en-us/library/gg398627.aspx`
- The *Import-CsConfiguration* post at `http://technet.microsoft.com/en-us/library/gg398800.aspx`

User database

Restoring core services in Lync is one thing, but the users will certainly complain if they discover that all of their saved contacts or conference information is lost. It is not *critical* to restore the user database, but it's highly recommended to have a backup of the user data. This is not only intended for disaster recovery but also to help single users if their information is corrupted.

Getting ready

PowerShell is the only tool available to perform this task, running in the elevated prompt. The commands listed here will fail unless we provide information about where the user databases are stored. Before running any of these commands, run `Get-CsService -UserDatabase` and take note of the listed database(s), as shown in the following screenshot:

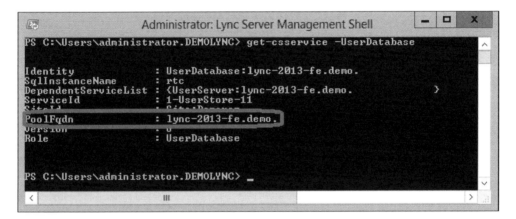

How to do it...

The following are the three steps to back up the user data and (or) conference directory:

1. Start PowerShell in the elevated administrative mode.
2. For each of the user database in your topology, type this:

   `Export-CsUserData -PoolFqdn "mydatabaseserver.domain.com"`
   `   -FileName "C:\Backup\BackUserDataFromMyDatabaseServer.zip"`

3. Save these files in a secure location.

Chapter 10

> In large deployments, it might be wise to have the name of the targeted database server in the filename for easier restore.

Restoring is as simple as the backup, as long as we target the correct user database (the backend server where the users are homed). It would not be of much help for the user if the data was restored to the wrong pool or server.

The following are the two steps to restore user data:

1. Start PowerShell in the elevated administrative mode.
2. For the selected user database, type this (this must be repeated for each user database in the topology we have to restore):

   ```
   Import-CsUserData -PoolFqdn "mydatabaseserver.domain.com" ´
   -FileName "C:\Backup\BackUserDataFromMyDatabaseServer.zip"
   ```

There's more...

Although the preceding examples show how to back up and restore the entire user database, the `Export-CsUserData` and `Import-CsUserData` commands are much more flexible. They can easily be used to export and import single users and/or conference directories. This is done using either `-userfilter` or `-confdirectoryfilter`.

By adding the following to the end of the `Export-CsUserData` string shown previously, `-userfilter myuser@lyncdomain.com`, the command exports user data only for the specified user. This will allow an administrator to go through the specified users' contacts; possibly remove, add, or change information; and then import the information back into the database.

By adding the same string to the end of the import string shown previously, the single user is restored as described earlier. This allows for a simple restore of a single user who might have been deleted, or the Lync account might be corrupted:

```
Import-CsUserData -PoolFqdn "mydatabaseserver.domain.com" 1.
-FileName "C:\Backup\BackUserDataFromMyDatabaseServer.zip" -userfilter
myuser@lyncdomain.com
```

> If, for some reason, the user database has been moved from one version to another version of Lync, we have the `Convert-CsUserData` command to change the file format between XML (Lync 2010) and ZIP (Lync 2013). The command for this is as follows:

```
Convert-CsUserData -InputFile "C:\Backup\
BackUserDataFromMyDatabaseServer.zip" -OutputFile " C:\Backup\
converteuserdatato2010.xml" -TargetVersion Lync2010 -UserFilter "
myuser@lyncdomain.com"
```

> This command will convert the data for a user who has been moved from Lync 2013 to Lync 2010 and then needs to be restored into Lync 2010 with the appropriate tools (`DBImpExp.exe`).

The users must be offline when you run the `Import-CsUserData` command.

> There is a different command to update the database if a user is online at the time of restoring:
>
> ```
> Update-CsUserData -Filename "C:\Backup\
> BackUserDataFromMyDatabaseServer.zip " -UserFilter
> myuser@lyncdomain.com
> ```
>
> If this is a large deployment with different pools, use the `-routinggroupfilter` switch to identify the correct user pool.
>
> Note that the update command does not write to the database but through the fabric.

See also

- The *Export-csuserdata* post at http://technet.microsoft.com/en-us/library/jj204897.aspx
- The *Convert-csuserdata* post at http://technet.microsoft.com/en-us/library/jj205337.aspx
- The *Update-csuserdata* post at http://technet.microsoft.com/en-us/library/jj205358.aspx
- The *Import-csuserdata* post at http://technet.microsoft.com/en-us/library/jj205373.aspx

Chapter 10

Persistent Chat database

The Persistent Chat Server role is optional, and not all deployments include this role. If the role is not implemented, there is really nothing to worry about here.

In the event that the Persistent Chat Server role is deployed, it could be on a separate pool and have a different set of databases than the rest of the Lync deployment. This will vary between the Standard Edition (collocated) and Enterprise Edition deployments.
The backup and restore is done with their own set of backup commands and must be performed using PowerShell.

The Persistent Chat database and information is *not* included in the `Export-CsConfiguration` or `Get-CsTopology` command (except for the name of the pools, database, and core information).

> Run the following PowerShell command to verify that the Persistent Chat is deployed:
> `Get-CsService -PersistentChatDatabase`

Getting ready

This task can only be performed in PowerShell and must be done in the elevated administrative mode.

To get a list of the servers and instances in use by the deployment, run this:

`Get-CsService -PersistentChatDatabase | Select-Object poolfqdn, sqlinstancename | fl`

How to do it...

The `Export-CsPersistentChatData` and `Import-CsPersistentChatData` commands are primarily for migration purposes, but why not use them as an extra backup?

Managing Lync 2013 Backup and Restore

The following are the three steps to back up the Persistent Chat databases:

1. Start PowerShell in the elevated administrative mode.
2. For each server returned by the `get` command, run this:

    ```
    Export-CsPersistentChatData -DBInstance´ sql_srv.mydomain.com\
    mypreschatinst -FileName´ "c:\backup\PersistantChatDBx.zip
    ```

3. Store those files in a secure location.

Name the export ZIP files in a such a way that you can recognize where they belong when a restore is in need.

If the deployment is configured with pool pairing, only the master database is required for backup. To verify this, run the following command:

`get-csdatabasemirrorstate -poolfqdn mychatpool.mydomain.com -databasetype persistentchat`

The following are the two steps to restore data:

1. Start PowerShell in the elevated mode.
2. For each server in the deployment where a restore is needed, run the following command:

    ```
    Import-CsPersistentChatData -DBInstance´ sql_srv.mydomain.com\
    mypreschatinst -FileName´ "c:\backup\PersistantChatDBx.zip"
    ```

The topology must be deployed with servers, database servers, and instances, before this task can begin.

There's more...

The `Export-CsPersistentChatData` command can be used to make a more granular export/backup. There are a couple of switches to control the output. These can be very useful to control the amount of data exported. The switches are as follows:

- `- Level`: This is used to control what type of data we export. The values that are allowed are `All` (default), `User`, `Category`, `RoomDirectory`, and `Content`.
- `- Scope`: This is used to control what category to export.

- *Startdate*: This is used to control how much *old* data to export. Not all data from the beginning of time is relevant when doing backups. This switch can only be used when `Level` is set to `RoomDirectory`.

 Running the `Export-CsPersistentChatData` command without specifying any of these additional switches will create a backup *from the beginning of time*. If this is undesirable, the export job should be split into several jobs using the `Level` switch. However, in a large deployment, this can easily get very complex and hard to control. Keep it as simple as possible.

See also

- The *Export-CsPersistentChatData* post at http://technet.microsoft.com/en-us/library/jj205378.aspx
- The *Import-CsPersistentChatData* post at http://technet.microsoft.com/en-us/library/jj204709.aspx

The Location Information LIS database

Creating a fully functional Enterprise Voice Enhanced E-9-1-1 configuration is a lot of work in large deployments. The LIS database is a database used to identify the location of a call coming from somewhere in the campus and going out to a certified emergency services provider.

Losing all of this information in the event of a system failure can be frustrating. As networks grow and change over time, this information will change as well. A regular backup of this information is highly recommended. This information is *not* included in the backup made with `Export-CsConfiguration` or `Get-CsTopology`.

Getting ready

PowerShell is the best tool to perform these tasks. This task requires an account that is a member of **RTCUniversalServerAdmins**.

Managing Lync 2013 Backup and Restore

How to do it...

The following are the steps to back up your LIS configuration:

1. Start PowerShell in the elevated administrative mode.
2. Run the following command:

   `Export-CsLisConfiguration -FileName C:\Backup\LisConfig.bak`

3. Store the file in a secure location.

> The LIS configuration is stored and replicated in the backend database store, and the export/import command is run only once, without specifying the FQDN of the Lync Front End pool.

Follow these steps to restore your LIS configuration:

1. Start PowerShell in elevated administrative mode.
2. Run the following command:

   `Import-CsLisConfiguration -FileName C:\Backup\LisConfig.bak`

3. Then run the following command:

   `Publish-CsLisConfiguration`

> If this is a partial restore, run to correct some errors, be aware it will not rewrite the entire LIS Database. The `Import` command will add missing entries, and it will skip existing entries. It *may* create erroneous locations. It is wise to test and verify the functionality after a partial import (a total disaster recovery will not have this issue, as there is no data in the store).

There's more...

After a restore, there are two very handy commands for verification and testing. One can give a useful insight into the published configuration, the other will test the database based on the input in the command.

To get a good view of the E-9-1-1 configuration, use the following command:

`Debug-CsLisConfiguration | ft -wrap`

Chapter 10

The `-wrap` switch is used to get all the information in the window. Without the `-wrap` switch, the output will be wide (lots of scrolling from side to side, in the absence of a very wide screen).

The `Debug-cslisconfiguration` command will by default create an XML output (not all `get-` commands will), which is why we pipe it to format-table (ft).

A great way to test the configuration is to enable **CsHealthMonitoringConfiguration** and run the `Test-CsLisConfiguration` command with several scenarios.

Setting up **CsHealthMonitoringConfiguration** is not a part of the scope for this chapter, but the following Technet article explains the process: http://technet.microsoft.com/en-us/library/gg398718.aspx.

With the Health Monitoring set up, the following command can be run on any computer running PowerShell with the Lync Module:

`Test-CsLisConfiguration -TargetFqdn mylyncregistrarpool.domain.com -Subnet 10.0.42.0`

If the Health Monitoring is not available, the following commands should be run on a computer with a server certificate (typically on a Front End Server):

`Test-CsLisConfiguration -TargetFqdn mylyncregistrarpool.domain.com -Subnet 10.0.42.0 -UserSipAddress sip:myuser@mysipdomain.com`

The command can still be run on a non-Lync Server/computer, but then it will require credentials to perform the task, like this:

`$logOn = Get-Credential`

`Test-CsLisConfiguration -TargetFqdn mylyncregistrarpool.domain.com´`
  `-Subnet 10.0.42.0 -UserSipAddress sip:myuser@mysipdomain.com´`
    `-UserCredential $logOn`

The first line of code will prompt you for the username and password for the tested subject (identified with the sipURI in the following line).

See also

- The *Export-cslisdatabase* post at `http://technet.microsoft.com/en-us/library/gg398539.aspx`
- The *Import-CsLisDatabase* post at `http://technet.microsoft.com/en-us/library/gg398380.aspx`
- The *Publish-CsLisconfiguration* post at `http://technet.microsoft.com/en-us/library/gg398364.aspx`
- The *Debug-CsLisconfiguration* post at `http://technet.microsoft.com/en-us/library/gg398710.aspx`
- The *Test-CsLisconfiguration* post at `http://technet.microsoft.com/en-us/library/gg398497.aspx`

The Response Group Services configuration

Response Group Services is yet another part of Lync, not covered with the export of the configuration or topology. For this task there is another set of export/import commands available in PowerShell. There are no other tools for this yet.

The export and import commands will take care of the following objects in your deployment: queues, agent groups, workflows, holiday sets and business hours, audio files and service configuration settings.

Getting ready

A user with **RTCUniversalServerAdmins** rights must run these tasks in an elevated PowerShell prompt.

Before running the command, we need to identify the pools running the application service as follows:

```
Get-CsService -ApplicationServer
```

How to do it...

The following are the steps to back up your Response Group configuration:

1. Start PowerShell in elevated mode.
2. For each application pool, run the following command:
   ```
   Export-CsRgsConfiguration -Source
   ApplicationServer:mylyncapppool.domain.com"
   -FileName "C:\Backup\mylyncapppoolRgs.zip"
   ```
3. Store the files in a secure location.

 When there are several pools in your deployment, naming the backup file according to the pool it came from is a good idea.

The following are the steps to restore your configuration:

1. Start PowerShell in the elevated mode.
2. For each application pool, run the following command:
   ```
   Import-CsRgsConfiguration -Destination
   "ApplicationServer:mylyncapppool.domain.com"
   -FileName "C:\Backup\mylyncapppool _Rgs.zip"
   ```
3. Restart the Response Group services.

There's more...

Note that these commands are intended for Lync 2013. If there are any Lync 2010 pools that run RGS, there is a separate tool in **Lync 2010 Resource Kit Tools** for the job.

In a migration scenario, we should be moving the RGS configuration from Lync 2010 to Lync 2013 rather than export/import. This is done with the `move-CsRgsConfiguration` command after all the RGS users have been migrated to Lync 2013. Do not move the configurations from 2010 to 2013 if there are still RGS users left in the 2010 pool.

See also

- The *Export-CsRgsConfiguration* post at `http://technet.microsoft.com/en-us/library/jj205011.aspx`
- The *Import-CsRgsConfiguration* post at `http://technet.microsoft.com/en-us/library/jj205245.aspx`

Certificates

Certificates are important in the Lync infrastructure. They are used to authenticate, sign, and secure communications. It is important to have a strategy on how to restore the Lync functionality where certificates are highly indispensable.

In the *Don't forget the infrastructure – the greater recovery plan* section, the internal PKI is listed. There is no point in backing up or restoring certificates issued from the internal CA, unless this is on the disaster recovery plan for the company. This section is worth reading if the PKI is on the restore list or if Lync is the only service we have to restore.

> Some say that it's faster or quicker just to request and assign new certificates, and they will possibly skip the entire sections.
> However, one might not always be in a position where the online CA is available; this is why a backup is recommended.

Getting ready

There is no way to create a backup of the systems certificates unless the right **allow export of private key** option has been set during the deployment phase and/or the issuing CA allows it (most CAs do).

When requesting the certificate in the Lync deployment wizard, the **Mark the certificate's private key as exportable** option must be selected, as shown in the following screenshot:

If the request is done through the Certificate MMC snap-in, the corresponding setting would be the **Make private key exportable** option in the **Private Key** tab of a custom request, as shown in the following screenshot:

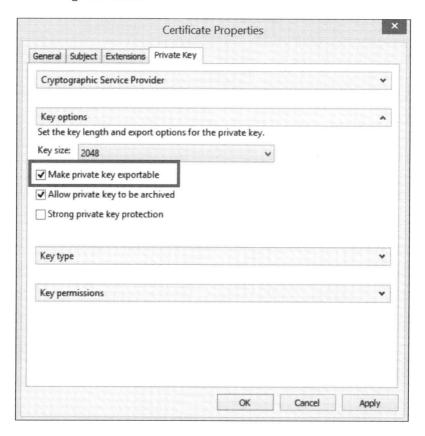

The Lync Server Deployment tool does not have any option or settings to export (take a backup) a certificate. This task requires the use of the Certificate MMC snap-in (computer store).

On every windows server where there is a certificate we want to back up, perform the following steps to load MMC:

1. Open the **Start** menu.
2. Select **Run**.
3. Type mmc and press *Enter*.
4. Press *Ctrl +M* (add snap-in).

5. Select **Certificate**, click on **Add**, and select **Computer account**, as shown in the following screenshot:

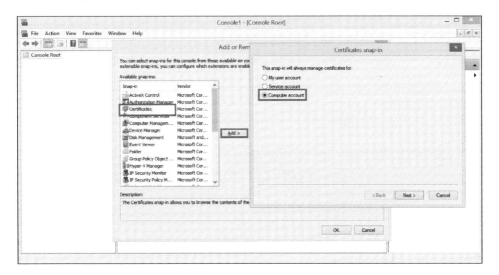

6. Click on **Nex**t and then on **Finish** (leave the default of local computer and then click on **OK**).

> It is not uncommon to create certificate requests for hardware load balancers on a Windows server, using either MMC or the Lync Server Deployment tool. If this is the case, the machine on which the request was created should have a copy of the certificate, and the backup can be made from there (it is recommended that you create the backup when requesting and assigning the certificates). This guide only covers Windows Server tools.

How to do it...

The following are the steps to create the backup on a Windows computer through the MMC snap-in:

1. Expand the **Console root**, **Certificate (Local computer)**, and **Personal** options in the left pane.
2. Select **Certificates**.
3. Locate the desired certificate in the main pane and right-click on it.

> Identify the correct certificates by running the following PowerShell command (most likely, there two different certificates: one for OAuth and another for the rest of the services):
>
> `Get-CsCertificate | ft -Property use,subject -AutoSize`

4. Select **All tasks** and then **Export** (this launches a wizard).
5. Click on **Next**.
6. Select **Yes, export private key**, as shown in the following screenshot:

7. Click on **Next**.
8. Keep the wizard's default values and click on **Next**.
9. Select **Password**. Then, create a password (and document it) to use for the certificate's private protection.
10. Click on **Next**.

11. Select a path and a filename. Then, click **Next**.
12. Click on **Finish** and then on **OK**.
13. Save the *.pfx file in a secured location.

Follow these steps to restore a certificate to a Windows server using MMC:

1. Expand the **Console root**, **Certificate (Local computer)**, and **Personal** options in the left pane.
2. Right-click on **Certificates**.
3. Select **All tasks** and **Import**.
4. Select **Next**.
5. Browse to the PFX file that contains the certificate in question (remember to change the type to PFX; by default, the wizard browses for .cer).
6. Click on **Next**.
7. Enter the password you created at the export stage, select **Make this key exportable** (if a re-export is needed), and then click on **Next**.
8. Leave the default (import into personal store).
9. Click on **Next**.
10. Click on **Finish** and then on **OK**.

> This task only described the import job. The certificate must be reassigned to the Lync services in **Deploy Lync Server 2013 Wizard** in the same way in that the administrator did the initial deployment.

11. Launch **Lync Server 2013 – Deployment Wizard**.
12. Select **Install or Update Lync Server System**.
13. Select **Run** or **Run again** (depending on the current system status) under **Step 3: Request, Install or Assign Certificates**.
14. Select the service for the reassignment and select **Assign**.
15. Click on **Next**.
16. Select the correct certificate and then click on **Next**.
17. Click on **Next** again.
18. Click on **Finish**.
19. Restart services.

There's more...

Not all certificates on all servers have to be backed up. Why? In simple deployments, it might be just as easy and quick to restore certificates by re-requesting and reissuing certificates from the Internal CA where the original certificates originate from.
Make a risk assessment and write down the decision regarding internal certificates (and/or procedures) in the recovery plan.

External (public) certificates should always be backed up and stored for an emergency, as the external CA might not be available at the time of recovery.

As certificates must be exported from each server individually, the procedure described earlier can be slow and and take some time.

There is one way to export certificates with a script or in a PowerShell session. The following commands show you how to do it:

```
dir cert:\localmachine\my | '
    Where-Object { $_.HasPrivateKey '
    -and $_.PrivateKey.CspKeyContainerInfo.Exportable } | '
    Foreach-Object { [system.IO.file]::WriteAllBytes( '
        "C:\backup\$($_.thumbprint).pfx", '
        ($_.Export('PFX', 'password')) ) }
```

The first line (`dir`) is to set the path of the store. The second command (`where-object`) is to select certificates for the export process (the private key is marked for export).

The `foreach` command is to repeat the export for eligible (Lync and other) certificates, presuming that `C:\backup` already exists.

Backend databases

Backing up databases used to be challenging or had to be done with SQL management tools. Thanks to new PowerShell commands shipped with SQL server 2012 and SQL server 2012 Express (the latter shipped with Lync 2013), backing up and restoring has become a lot easier.

The script examples given here presume that the Lync backend databases are on an isolated instance (colocation core services and monitoring databases). If not, it would be wise to read up on the list of databases to take backup of at http://technet.microsoft.com/en-us/library/gg398479.aspx. Then, adjust the following scripted examples to back up or restore only the databases needed for Lync.

> QOE and CDR databases can grow dramatically in large deployments and can take some time to back up and restore. Consider whether a SQL backup for these databases are vital or not.

Getting ready

The methods described for Lync Standard and Enterprise are slightly different. Standard Edition backup is done on the Lync Server, whereas a backup of an Enterprise Edition should be done on the SQL server.

Run the `Get-CsService -CentralManagementDatabase` command to identify the server and instance where the Lync Server has its databases.

How to do it...

The following are the five steps to back up your databases on a Standard Edition server:

1. Launch PowerShell in the elevated administrative mode on the Front End Server for Standard Edition or on the SQL server for Enterprise Editions.
2. Type `Import-Module SQLPS -DisableNameChecking` (to import the SQL commands if the SQL server is not running PowerShell v3).
3. Type `cd SQLSERVER:\SQL\Frontenserver.mydomain.com\SQLInstance\Databases` (the server and instance information was gathered in the *Getting ready* section).
4. Type the following `foreach` loop (no line breaks; they are here for viewing). The `foreach` command here will look at all the databases in the instance (except `tempdb`) and then back them up to individual files:

   ```
   foreach($database in (Get-ChildItem -name -force)) {'
   $dbName = $database'
   $bakFile = "c:\backup\" + $dbName + "_full_.bak"'
   If($dbName -ne "tempdb"){'
   Backup-SqlDatabase -Database $dbName -BackupFile $bakFile -Initialize}}
   ```

5. Store the files in a secure location.

How to restore...

Restoring is not much different from backing up, and here are the four easy steps to restore your databases:

1. Launch PowerShell in the elevated mode on the Front-End Server for Standard Edition or on the SQL server for Enterprise Editions.
2. Type `Import-Module SQLPS -DisableNameChecking` (to import the SQL commands).
3. Type `cd SQLSERVER:\SQL\Frontend?server.mydomain.com\SQLInstance\Databases` (the server and instance information was gathered in the *Getting ready* section).
4. For each database to restore, type the following command (replace `rtc` with the correct database):

   `Restore-SqlDatabase -ServerInstance SQLInstance -Database rtc -BackupFile "c:\Backup\Rtc_full_.bak" -ReplaceDatabase`

> You will need to restore each database individually, and if the `-Replace` switch is used, it would overwrite everything in the target location.

Voice dial plans, policies, and settings

Strictly speaking, the **Voice Dial Plan**, **Voice Policies**, and other voice-related configuration are all backed up in the **CsConfiguration** and **Topology Information** backups. However, there are times when we need to restore only portions of the configuration and settings by comparing settings in a backup with a system in production (especially if something suddenly got broken).

That's why we have chosen to include this set of voice-related information, which can be exported separately. There is no real need for these exports unless you need the flexibility that we described previously.

These exports cannot be natively imported back into Lync, but with the information provided within them, it suddenly becomes easier to recreate single settings and policies.

Getting ready

1. Load PowerShell in the elevated administrative mode.

 In order to save this as files, we need to use `export-clixml`.

How to do it...

1. Run the following commands in PowerShell:

   ```
   Get-CsDialPlan | Export-Clixml -path c:\backup\DialPlan.xml
   Get-CsVoicePolicy | Export-Clixml -path c:\backup\VoicePolicy.xml
   Get-CsVoiceRoute | Export-Clixml -path c:\backup\VoiceRoute.xml
   Get-CsPstnUsage | Export-Clixml -path c:\backup\PstnUsage.xml
   Get-CsVoiceConfiguration | Export-Clixml -path c:\backup\VoiceConfiguration.xml
   Get-CsTrunkConfiguration | Export-Clixml -path c:\backup\TrunkConfiguration.xml
   ```

2. Store the files for documentation.

There's more...

The XML files like this can be hard to read. However, PowerShell has the tools to help us out. Here are a few examples to choose from:

The first way is using **Out-Gridview.** Import the XML file and pipe it to the gridview as follows:

```
Import-Clixml 'C:\Backup\TrunkConfiguration.xml' | Out-GridView
```

Managing Lync 2013 Backup and Restore

This will provide a graphical interface to look at for comparison between the live environment and the exported information, as shown in the following screenshot:

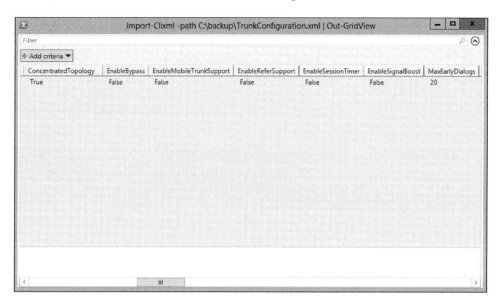

There is a lot of information to cover, and we need to keep scrolling to the right to see more of the settings applied to the trunk, as shown in the following screenshot:

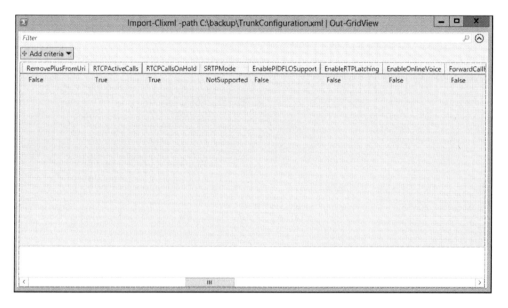

Depending on the screen size, we might have to scroll three, four, or even five pages to get to the end, as shown in the following screenshot:

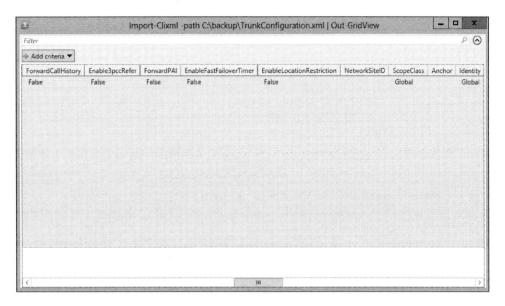

The final page includes some of the other pages and (probably, most important) any translation rules applied to the trunk, as shown in the following screenshot:

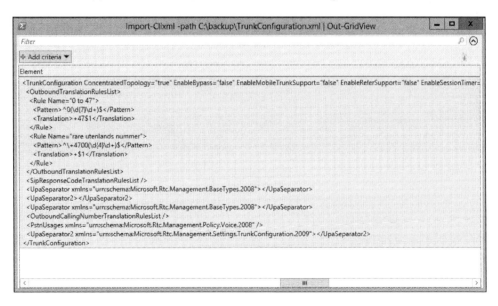

Managing Lync 2013 Backup and Restore

Another way to compare the old and new information is to import the file. Then, pipe the import to see the values the same way we do with `get` commands, as shown here:

`$ImportToCompare = Import-Clixml -path    c:\backup\TrunkConfiguration.xml`

Then, run the following command:

`$ImportToCompare`

This will give the same output as the `Get-CsTrunkConfiguration` command does, only from the stored information. The `$inputToCompare` command can then be manipulated any way we want with the `Get-CsTrunkConfiguration` command. We can format-list or format-table or select chosen properties for comparison, as shown in the following screenshot:

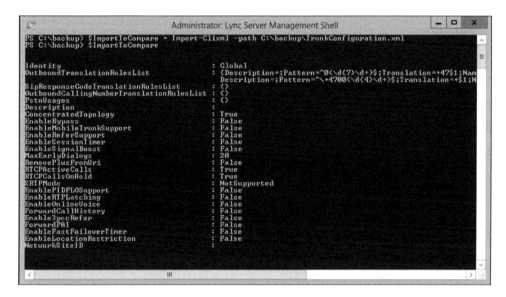

File services

Performing a back up of the file store has to be one of the easiest parts of the Lync infrastructure. If the deployment is an Enterprise version, odds are that the file store is already on a file server that is regularly backed up. However, if there are Standard Edition servers in the deployment, the file store is probably on the Lync Server, and there might not be any file back up on these servers.

Chapter 10

The best solution is to have a backup agent configured and running on each server where there is a file store configured. However, be aware of resources when planning for this and consider only off-peak backup times.

The *backup* solution described here will just show you how to make an **xcopy** of the files to a second location for safekeeping.

Getting ready

Start by finding out where the file stores are and whether or not a separate backup of these files are necessary. This task is easily done in PowerShell with the Lync module loaded type: `get-csservice -filestore`.

On any computer/server where the copy of the filestore is to be stored, decide in which location the files will be copied, for example, `C:\LyncFileBackUp`.

How to do it...

The following are the three steps to create a backup:

1. On computer/server where the copy of the filestore is to be stored, start CMD.
2. Map a drive to the file store as follows:
   `net use g: \\lyncfileserver\sharedfolder`
3. Then, run the following command:
   `Xcopy g:\*.* c:\LyncFileBackUp /E /I /Y /H /C`

The following are the three steps to restore:

1. On computer/server where the copy of the filestore is stored, start CMD.
2. Map a drive to the file store:
   `net use g: \\lyncfileserver\sharedfolder`
3. Then, run the following command:
   `Xcopy c:\LyncFileBackUp\*.* g:\ /E /I /Y /H /C`

Managing Lync 2013 Backup and Restore

There's more...

There are certain folders in the file store that we don't always need. These folders are the `DeviceUpdateStore` and `ABFiles` folders.

The `ABFiles` folder contains the address book files. These are easily recreated at the point of restore. Simply type `Update-CsAddressBook`, and the files are recreated based on your Active Directory data.

The `DeviceUpdateStore` folder can contain a large amount of old data, as old updates are not removed when new ones are imported. A good approach to this could be to exclude the directory, download the latest update pack from TechNet or other specific vendor updates from their sites at the time of restore, and then just install it.

 The Lync 2013 update page at http://technet.microsoft.com/en-us/office/dn788954.aspx is the best place to look for current updates.

The following are the steps to exclude these folders at the time of backup:

1. On computer/server where the copy of the file store is to be stored, start CMD.
2. Map a drive to the file store as follows:

   `"net use g: \\lyncfileserver\sharedfolder"`

3. Create an empty `.txt` file in the computer, for example, `C:\temp\exclude.txt`.
4. Populate this file with the following two words on separate lines:
   - `ABFiles`
   - `DeviceUpdateStore`
5. Save the file.
6. Then, run the following command:

   `Xcopy g:\*.* c:\LyncFileBackUp /E /I /Y /H /C /EXCLUDE:c:\temp\exclude.txt`

Don't forget the infrastructure – the greater recovery plan

This section is, strictly speaking, not a *how to* section but rather a huge checklist of all the other components one should consider in a complete backup/restore scenario for an organization's Universal Communications solution.

Lync does not stand alone in the infrastructure; it relies on many other components, and it would be fatal to neglect this. Restoring only Lync will not get you anywhere near a functional system.

In the final paragraphs, we'll try to summarize some of the other components we must pay attention to when considering the disaster recovery plan.

Let's talk about some of the dependencies in the infrastructure that are not covered by the procedures described so far. Just to make things clear, these are as important as the Lync infrastructure itself. In order to have a full and healthy backup, consider including the following elements described hereafter in your disaster recovery plan.

This section will not describe how to make backups or do restores of other systems, as it is beyond the scope of this book. It is here for reference and to give Lync administrator s a clearer understanding of the complexity of a disaster recovery.

Make a checklist, and make sure that everything is covered.

Active Directory

As Lync is installed or upgraded, the schema is extended with attributes. These attributes have Lync-related information in them for each Lync-enabled user or contact (or other objects). Without a proper backup of AD, there is no way to recover from a total disaster, not without starting from scratch anyway.

Active Directory is the place where new servers look for information about the location of the Central Management Store.

DNS (Internal and external)

Think of all **A-records**, **CNAME-records**, and **SRV-records** of servers, gateways, pools, and simple URLs. Some of them can be autocreated when servers are added to the domain. However, a lot of them are added manually.

Make sure that you have at least a proper documentation of the internal and external DNS zone, so that the configuration can easily be recreated.

Depending on the overall backup scheme, this backup might be covered by backing up AD if the internal zones are integrated in AD.

DHCP

Devices such as Lync Phone Editions use **Dynamic Host Configuration Protocol** (**DHCP**), and this is sometimes overlooked when it comes to backup. Within DHCP is the configuration for these devices on how to connect to the Lync pool. Make regular backups of the DHCP configuration as well. Call Admission Control; otherwise, E-9-1-1 might be implemented, and it would clearly give the administrator a real headache if the subnets are shifted after a recovery.

PKI (Internal Certificate Authority) Infrastructure

A proper backup and recovery plan of a **public key infrastructure** (**PKI**) is a complete chapter in itself. I just want to make you aware of all the certificates used in the internal deployment to create service and trusts; Lync deployments are dependent on the PKI infrastructure. With the loss of the PKI, the entire PKI will have to be recreated, and all certificates have to be recreated and deployed.

File servers (not the shares, but the actual servers)

The Lync file share is covered earlier in this chapter, but the server itself also needs to be restored if disaster strikes. This can be easily done in a virtualized environment with a snapshot every now and then.

The SQL Server

Again, the different Lync databases are already covered, but the server running SQL software should also be on the list for disaster recovery. How do we restore databases if the SQL server should fail? One way of mitigating this risk is using clustering or mirroring on the backend.

The Lync Server backup

To save time in the event of a disaster, it would help to have the capability to restore the Lync Servers without installing everything from scratch. Think of the actual OS of the physical or virtual machines that run the services. If Lync is in a virtualized environment, snapshots are a great starting point.

Exchange backup

Lync depends on Exchange in the following ways:

- Users have their mail accounts in Exchange
- Users might have their conversation history stored in their mailbox
- Their presence is synchronized from their calendar

- Contacts are stored in the Unified Contact store
- Archiving is a component that can be collocated with Exchange 2013
- The Unified Messaging service might be in use

Pay close attention to this in the recovery plan. Think of servers, roles, configurations, databases, and mail flows.

Firewall configurations

Firewalls are also an important piece of the puzzle. Some are only external, others are internal or a mix of the two. Nevertheless, we need to be able to recreate firewalls and the rules in effect for all clients and servers. The recovery plan should include backups and separate documentation of your firewalls.

Router and switch configuration

The network is as important as anything else for the Lync installation. A disaster recovery plan should have a detailed description of how to back up and restore network connectivity exactly as it was, as soon as possible.

Again, think how bad it would be for elements such as QOS, CAC, and 9-1-1 if things got lost or recreated in the wrong way. Bad quality calls could start happening, or calls could erroneously be denied. A 9-1-1 call might be sending the wrong information to the emergency services and dispatch rescuers to the wrong address.

As with the firewalls, make sure that you have backups and a detailed documentation.

Reverse proxy

A reverse proxy is a required component to enable and allow features such as mobility, remote login, and federation to work properly. Remember to back up these services as well.

Miscellaneous

To complete the disaster recovery plan, have a clear idea of where to find proper installation media for Windows Server, SQL, Lync, clients, and so on. I know it can be downloaded, but we might face Internet connectivity issues at the point of recovery. Have it all downloaded and updated with the current releases and have them included in the backup.

As the recovery plan is created, try to think of other elements in the organizations infrastructure that are not mentioned earlier. There might be some kind of a third-party application or solution that the organization depends on. It should all be there on the checklist, next to all of the items described earlier.

11
Controlling Your Network – A Quick Drill into QoS and CAC

In this chapter, we will cover the following topics:

- Gathering data about your network
- Creating network bandwidth policies
- Adding networks to the topology
- Creating region links and routes
- Enabling CAC
- Preparing servers and clients for DSCP tagging
- Controlling/limiting the port ranges for traffic
- Media bypass

Controlling Your Network – A Quick Drill into QoS and CAC

Introduction

Many admins believe that they can implement Lync in their organization without implementing **Quality of Service** (**QoS**) and **Call Admission Control** (**CAC**). To some extent, this is true. If the organization is only implementing a **Presence** (**P**) and **Instant Messaging** (**IM**) solution with the odd **Peer to Peer** (**P2P**) audio/video session, it might seem like a huge and costly project to implement QoS and CAC for these services. IM and P do not rely on a real-time protocol for deliverance.

> QoS can be described as a technique used to identify certain types of traffic to give it priority throughout the network. CAC is about limiting how much bandwidth each call can utilize or how many calls can be there between the subnets in question.

However, not implementing a way to control the traffic can soon enough turn out to be a bad decision. Consider a scenario where Lync is implemented as an IM and P solution. However, over time, users become aware of the audio, video, and sharing capabilities within the product. Slowly, the use of real-time protocols grows until the network can no longer deliver at the required performance. Once the users start depending on audio and video, they will soon notice if a call sounds choppy or strange, or if their video breaks into pixels or freezes entirely.

If the users experience poor quality, there is a real chance that they will stop using the product in favor of other products that they experience fewer issues with (things such as going back to the old physical phone, skipping videos, and avoiding online meetings if a physical one is an option).

This is where QoS and CAC can help an organization. Implementing QoS and CAC are two ways of controlling the flow of traffic throughout the network.

> There is no point in implementing QoS or CAC within Lync, unless the rest of the infrastructure is configured accordingly. This, however, is beyond the scope of this chapter and book. QoS and CAC have to be implemented as a cooperative effort with those in control of the network (routers, switches, firewall, and other networking equipment).

This chapter will discuss how to implement QoS and CAC in your Lync deployment; it is not intended as an introduction to QoS and CAC in general. This chapter will go through a set of configurations needed to implement the core concept of QoS and CAC. In-depth information about all the inner workings is beyond the scope of this chapter.

Microsoft has provided a lot of good material that covers the topics of network, QoS, and CAC. The following are some recommended articles that you should be familiar with prior to starting this task:

- The *Planning your network* article at http://technet.microsoft.com/en-us/library/jj721883.aspx
- The *Networking guide* article at http://www.microsoft.com/en-us/download/details.aspx?id=39084
- The *Wi-Fi guidelines* article at http://www.microsoft.com/en-us/download/details.aspx?id=35401
- A list of certified equipment in general can be found at Microsoft's UCOIP at http://www.microsoft.com/en-us/download/details.aspx?id=35401

It is important to plan QoS and CAC implementation with the network administrators in the organization. Devices in the network will need to honor and keep markings in the path of a media stream; these can be (but not limited to) switches, routers, firewalls, and WAN links provided by external carriers.

Lync has a variety of codecs that it uses for different purposes, and the codec that the client selects can also have an impact on the **Quality of Experience** (**QoE**) for the end users.

> Internal calls would most likely select the adaptive codec RTAudio, which is a great codec to cover up minor errors in the network. However, it was never intended to be relied upon instead of implementing QoS and CAC over continuously congested network segments.

To mitigate network issues, Lync also has a **Forward Error Correction** (**FEC**) functionality, which, in a jammed network where clients experience loss of packets, can make the client send a copy of every packet (doubling the load) in order to compensate.

> Read the TechNet article *Network bandwidth requirements for media traffic in Lync Server 2013* to get a more comprehensive understanding of the different codecs and their load on the network at http://technet.microsoft.com/en-us/library/jj688118.aspx.

Only by implementing QoS and CAC can we control and guarantee the same codec selection and QoE for every call. The client will choose the "best" available codec based on how much bandwidth we allow through CAC, and configured properly, this will be the same every time.

Most of the Lync configuration for QoS and CAC can be done either in the Lync control panel (CSCP) or in PowerShell. The primary tool for the walkthroughs given here will be Lync CSCP. The reason for this is the flow of the configuration. Almost everything is done in the **Network Configuration** section, and it is performed left to right.

> The first two tabs in **Network Configuration** are called **Global** and **Location Profile**. The **Global** tab is an exception to the left-to-right rule, and will be covered much later in this chapter. The **Location Profile** tab is not within the scope of this chapter but rather an important piece of the E-911 deployment.

We will be going through the following tabs:

- **Global**: This is a tab that turns the CAC and media bypass on/off for the deployment.
- **Bandwidth Policy**: This is a tab where policies that control bandwidth usages are created and maintained. These policies can be used to limit both single sessions and the total number of sessions (basically by setting **Audio Limit** x times higher than **Audio session limit**, where x represents the number of sessions) to and from a **Site** with a policy applied (and/or videos as well).
- **Region**: This is a tab where the **Region** settings are configured. It's possible to allow media to flow through alternate paths (such as over the Internet and between edge servers) through the **Enable audio alternate paths** or **Enable audio alternate paths** options within the configuration of each region.
- **Site**: This is a tab where **Site** settings are configured. Link the site to a **Bandwidth Policy**, define which **Region** a **Site** belongs to, or set a **Location Policy** for E-911, if applicable.
- **Subnet**: This is a tab for all subnets in the deployment. A subnet is configured with a **SubnetID** (network address) and a **Mask**. It is also tied to a site here.
- **Region Link**: This is a tab where two regions are connected as possible paths as well as the given **Bandwith policy** between them.
- **Region Route**: This is a tab where actual media paths (routes) between **Regions** are configured by tying together **Region Links**. A media call can't traverse outside its defined Region, unless a Region Route between the sites exists.

Gathering data about your network

Gathering all the pieces of the puzzle and planning the implementation is the most important part of QoS and CAC implementation. The implementation of QoS and CAC itself is not very difficult, but without proper planning, it could lead to an even poorer QoE for the end user. A poorly planned and implemented QoS scheme could, in other words, have a negative impact on the performance of the applications.

Never rush through this part; wait a week or two with the implementation rather than do it the wrong way.

Getting ready

Try to figure out the best way to document the network environment where the Lync traffic will be introduced. Decide whether to create Visio drawings, create a separate database, keep the records in an Excel spreadsheet or Notepad, or do all of the above or whatever works best. In this recipe, we'll give a few examples of simple tables to reflect the data we talk about, but there are other ways to do this as well. As part of this task, you should be acquainted with more information from the following online resources:

- The *Lync Bandwidth Calculator* post at http://www.microsoft.com/en-us/download/details.aspx?id=19011
- The *Deploy Lync QOS* post at http://www.microsoft.com/en-us/download/details.aspx?id=12633
- Download and get familiar with the *Delivering Lync 2013 Real-Time Communications over Wi-Fi* post at http://www.microsoft.com/en-us/download/details.aspx?id=36494

> No matter which documentation alternative you decide on, it would be wise to select a method that allows data export to CSV files. Data can be entered manually into CSCP, but in a large deployment, an import from sources such as CSV files is really recommended.

How to do it...

There are a lot of important questions to be answered before the planning can begin. Some of the important ones are as follows:

- How many sites are there?
- Which subnets are configured in each site?
- How much bandwidth is there at each location in total?
- How much available bandwidth is there at each location, or how much can be reserved for real-time traffic?
- What kind of traffic is there to and from the locations today?
- How much of this traffic can be restricted?
- How many users are there?
- What kind of users are there in each site (referred to as profiles in the Lync Bandwidth Calculator LBC)?

Use all of the preceding information and the bandwidth calculator to decide on the answers to the following five important questions:

- How many concurrent audio calls should be allowed at any given time?
- How many concurrent video calls should be allowed at any given time?

> These first two questions might seem easy enough, but bear in mind that conferences might be using different codecs than an ordinary audio call. There might also be devices in the network that require certain codecs (and therefore a specific amount of KB available for sessions).

- What QoE or which audio codec will be allowed between sites?
- What QoE or which video codec will be allowed between sites?
- How many shared sessions should be planned for?

Based on all the information gathered, we could end up with a Visio that draws tables like the one shown as follows:

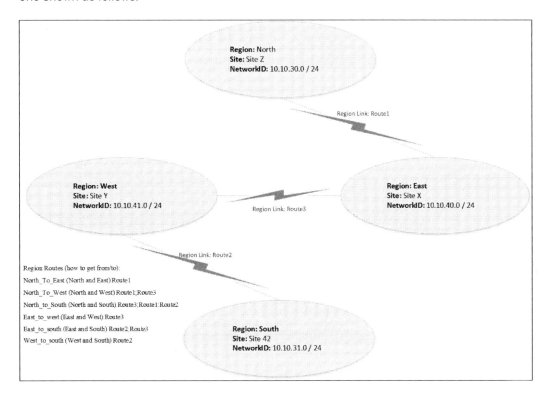

 This is a very simple setup; a Region can have many Sites within, and each Site can have many subnets associated.

The naming of table headers is based on the actual PowerShell command parameters used to import data, if applicable.

Controlling Your Network – A Quick Drill into QoS and CAC

The values for the **Differentiated Services Code Point** (**DSCP**) marking and port ranges are only examples, and are not to be implemented blindly. Do the calculations and avoid collisions based on the information provided in the white papers:

- **Network Bandwidth policies**

Identity	AudioBWLimit	AudioBWSessionLimit	VideoBWLimit	VideoBWSessionLimit	Description
Low	40	800 (20 Calls)	100	100 (Effectively 0 call)	For bad WAN links
Medium	60	2400 (40 Calls)	640	6400 (10 Calls)	Standard WAN links
High	80	8000 (100 Calls)	1500	75000 (50 Calls)	Best Quality

- **Regions**

Identity	CentralSite	AudioAlternatePath	VideoAlternatePath	Description
East	ROS	0	0	Test1
West	ROS	1	1	Test2
North	DAR	1	1	Test3
South	DAR	1	1	Test4

- **Network Sites**

Identity	NetworkRegionID	BWPolicyProfileID	Description
Site X	East	Medium	Remote X
Site Y	West	High	Remote Y
Site Z	North	Medium	Remote Z
Site 42	South	Low	VPN 42

Network Subnets

Identity	MaskBits	Description	NetworkSiteID
10.10.30.0	24	SkyMode	Site Z
10.10.31.0	24	EarthMode	Site 42
10.10.40.0	24	FireMode	Site X
10.10.41.0	24	WaterMode	Site Y

Region Links

Identity	NetworkRegionID1	NetworkRegionID2	BWPolicyProfileID
Route1	North	East	Medium
Route2	South	West	Medium
Route3	East	West	High

Region Routes

Identity	NetworkRegionID1	NetworkRegionID2	NetworkRegionLinkIDs
North_To_East	North	East	Route1
North_To_West	North	West	Route1;Route3
North_to_South	North	South	Route3;Route1;Route2
East_to_west	East	West	Route3
East_to_south	East	South	Route2;Route3
West_to_south	West	South	Route2

Table for QoS values

Type of traffic	DSCP Value
Audio	46 (EF)
Video	34 (AF41)
Signaling and app sharing	24 (CS3)
File transfer	18

Controlling Your Network – A Quick Drill into QoS and CAC

> **Table for port ranges**

Type of service	Port start	Port count (Number of ports)	Range =
Audio	50000	2000	50000 - 51999
Video	52000	2000	52000 - 53999
App sharing	54000	2000	54000 - 55999
File transfer	56000	2000	56000 - 57999
Edge service		10000	

Creating network bandwidth policies

It is a good thing to plan the entire CAC and QoS policy before beginning the configuration. Having done this, we can start adding the **bandwidth policies**. We could skip this and go directly to the **Region** tab, but once we get to the **Site** tab, we need to refer to these policies. Keep to the plan. Move left to right. Don't start this project unless the master QoS and CAC plan is complete.

> As an example, we will create three bandwidth policies in this walkthrough: Low, Medium, and High. The High policy will offer the best quality of audio and video. The Low policy will allow for a decent audio quality but no video. The Medium policy will be a balanced policy, allowing some video.

Getting ready

Begin this task only after carefully collecting data and when the planning of the entire QoS and CAC configuration is complete.

How to do it...

Proceed with the following steps for creating network bandwidth policies:

1. Launch the Lync Server Control Panel with an account that has the **RTCUniversalServerAdmins** rights, and navigate to **Bandwidth Policies** under the **Network Configuration** section.

2. Select **New** as shown in the following screenshot:

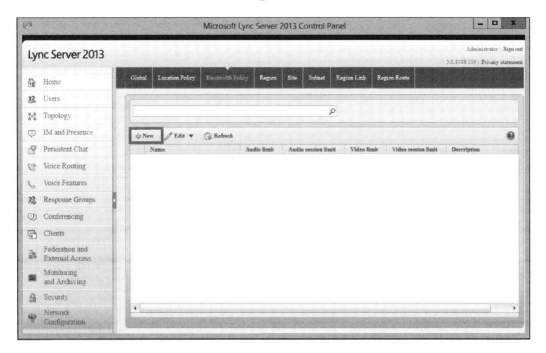

3. Enter the values desired for this policy (from the previous BW table).
4. **Name**: This is also known as Identity in PS.
5. **AudioBWLimit**: This is the total BW dedicated to audio.
6. **AudioBWSessionLimit**: This is the maximum BW per audio call.
7. **VideoBWLimit**: This is the total BW dedicated to videos.
8. **VideoBWSessionLimit**: This is the maximum BW per video call.
9. **Description**: This is not required but recommended.
10. Click on **Commit**.

Controlling Your Network – A Quick Drill into QoS and CAC

When this task is complete, the CSCP should look similar to the following screenshot:

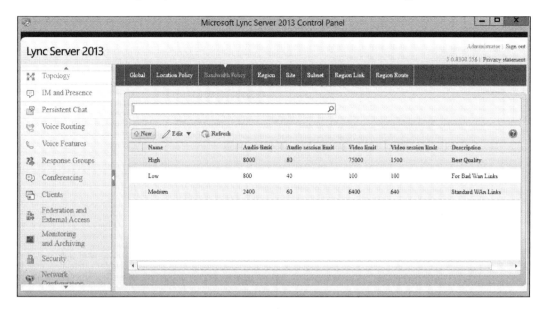

There's more...

This task can be completed in PowerShell as well, and if there is a call for many policies, PowerShell might be an easier and quicker way to apply these settings:

1. Create a CSV file with all the information required, as shown in the following screenshot (the header row is named according to the matching properties to be imported):

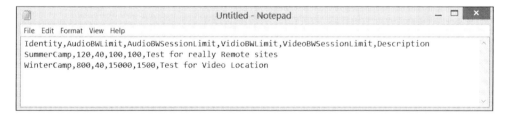

2. Save the file (the example script shows `C:\LyncScripts\QOSCAC.csv`).

3. Run the following PowerShell command to import and configure:

   ```
   Import-Csv C:\LyncScripts\QOSCAC.csv | foreach {
   New-CsNetworkBandwidthPolicyProfile -Identity $_.Identity`
    -AudioBWLimit $_.AudioBWLimit`
    -AudioBWSessionLimit $_.AudioBWSessionLimit`
    -VideoBWLimit $_.VideoBWLimit`
    -VideoBWSessionLimit $_.VideoBWSessionLimit`
    -Description $_.Description -Verbose
   }
   ```

See also

- The *New-CsNetworkBandwidthPolicyProfile* post at http://technet.microsoft.com/en-us/library/gg398675.aspx

Adding networks to the topology

Adding a network to the topology is a necessary step to implement CAC, not QoS. CAC uses the IP configuration of the client to decide which subnet and site the client belongs to. The information entered into Lync must match that of the client's IP configuration.

Servers can be *clients* too.

Supernetting two /24 subnets into one /23 subnet is not supported and will not work.

Subnets are added to sites, and sites are organized into regions. When entering this information into Lync, we have to create the regions first, then the sites, and finally the subnets.

Getting ready

Gather all the possible information on the layout of the infrastructure. Look at the previous tables for examples on how to document the deployment before entering data into Lync.

The tool that we'll use to show how it's done is CSCP, but look at the *There's more...* section for some PowerShell examples.

All of the following actions will require an account with the **RTCUniversalServerAdmins** permission.

Controlling Your Network – A Quick Drill into QoS and CAC

How to do it...

Proceed with the following steps to add a Region:

1. Launch the Lync CSCP and navigate to **Region | Network Configuration**.
2. Select **New**.
3. Add information related to **Name** and **Central site** (pick from the topology created for the organization), an alternate path (which allows a reroute), and a description, as shown in the following screenshot:

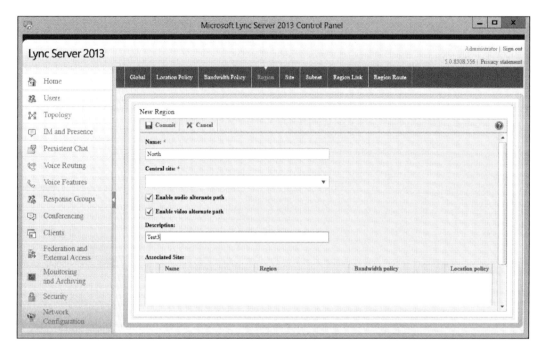

4. Select **Commit** to save.
5. The task is complete when all the regions have been added to the configuration.

The following are the steps to add a Site:

1. Launch the Lync Server Control Panel with a **CSAdministrator** account, and navigate to **Site | Network Configuration**.
2. Select **New**.
3. Add information related to **Name**, **Region**, **Bandwidth policy**, and **Description** (and **Location policy** if in use), as shown in the following screenshot:

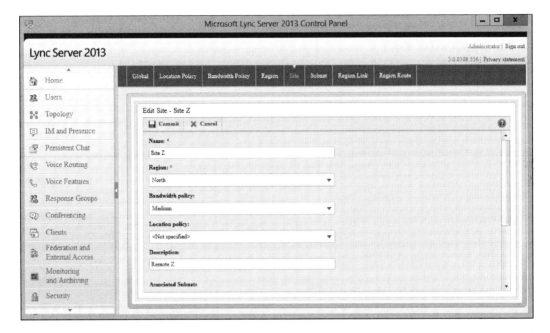

4. Select **Commit** to save.
5. The task is complete when all the regions have been added to the configuration.

The following are the five steps to add a subnet:

1. Launch the Lync Server Control Panel with a **CSAdministrator** account, and navigate to the **Subnet** option under the **Network Configuration** section.
2. Select **New**.

Controlling Your Network – A Quick Drill into QoS and CAC

3. Enter the information into **Subnet ID** and **Mask**, attach a site, and possibly add a useful description. Your screen will look similar to the one shown in the following screenshot:

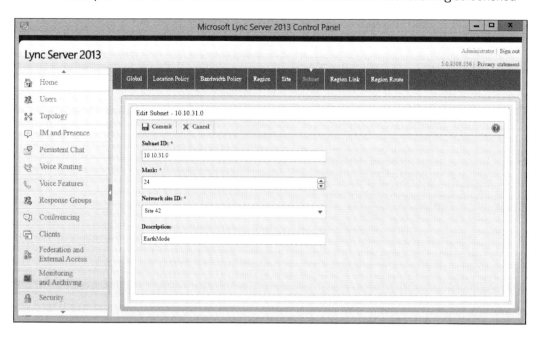

4. Select **Commit** to save.
5. The task is complete when all the regions have been added to the configuration.

There's more...

When a lot of information has to be entered into Lync, it makes much more sense to collect the information into CSV files and bulk import it into Lync.

Proceed with the following steps to add regions using CSV files:

1. Create a CSV file with all the information required, which looks similar to the following screenshot:

> 1 represents `$true` and 0 represents `$false`.

2. Save the file (the import example shows `C:\LyncScripts\Regions.csv`).
3. Run the following PowerShell command to import and configure:

```
Import-Csv C:\LyncScripts\regions.csv | foreach {
$_.AudioAlternatePath = [bool]($_.AudioAlternatePath -as [int])
$_.VideoAlternatePath = [bool]($_.VideoAlternatePath -as [int])
New-CsNetworkRegion -Identity $_.Identity`
 -CentralSite $_.CentralSite`
 -AudioAlternatePath $_.AudioAlternatePath`
 -VideoAlternatePath $_.VideoAlternatePath`
 -Description $_.Description
}
```

The following are the steps to add sites with the help of the CSV files:

1. Create a CSV file with all the information required. It should look similar to this:

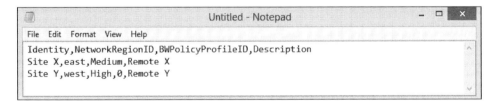

2. Save the file (the example shows `C:\LyncScripts\Sites.csv`).
3. Run the following PowerShell command to import and configure:

```
Import-Csv C:\LyncScripts\sites.csv | foreach {
New-CsNetworkSite -Identity $_.Identity`
 -NetworkRegionID $_.NetworkRegionID`
 -BWPolicyProfileID $_.BWPolicyProfileID`
 -Description $_.Description
}
```

Controlling Your Network – A Quick Drill into QoS and CAC

The following are the three steps to add subnets:

1. Create a CSV file with all the information required, which looks similar to this:

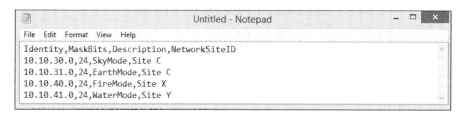

2. Save the file (the example shows `C:\LyncScripts\subnets.csv`).
3. Run the following PowerShell command to import and configure:

```
Import-Csv C:\LyncScripts\subnets.csv | foreach {
New-CsNetworkSubnet -Identity $_.identity`
 -MaskBits $_.MaskBits`
-Description $_.Description -NetworkSiteID $_.NetworkSiteID
}
```

See also

- The *New-CsNetworkRegion* post at http://technet.microsoft.com/en-us/library/gg425829.aspx
- The *New-CsNetworkSite* post at http://technet.microsoft.com/en-us/library/gg398365.aspx
- The *New-CsNetworkSubnet* post at http://technet.microsoft.com/en-us/library/gg398226.aspx

Creating region links and routes

Having created sites, subnets, and regions, we have still not told Lync how sites connect to each other. This is what region links and routes are for. Region links define bandwidth policies between sites, and region routes are used to connect sites by creating possible audio paths.

 When an audio path is chosen for the media stream, Lync will look at all the bandwidth policies in the path and select the most restrictive one for the call.

How to do it

The following are the five steps to add region links:

1. Launch the Lync Server Control Panel with an account with the **RTCUniversalServerAdmins** rights, and navigate to **Network Configuration | Region Link**.
2. Select **New**.
3. Enter a name for the link, select the two sites that will be connected through this link, and enter which **Bandwidth policy** should apply, as shown in the following screenshot:

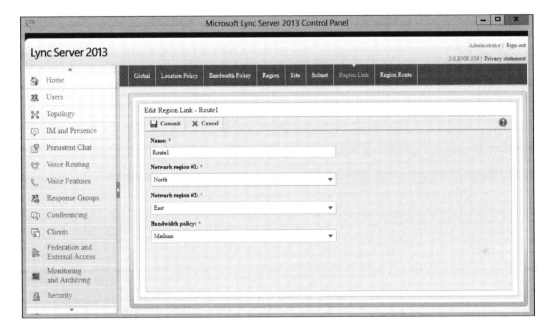

4. Press **Commit** to save.
5. This process is complete when all the links are added.

The following are the steps to add region routes:

1. Navigate to **Network Configuration | Region Route**.
2. Select **New** to add a new route.

Controlling Your Network – A Quick Drill into QoS and CAC

3. Add the two regions to be connected and the selected route, as shown in the following screenshot:

4. The task is complete when all the regions and their links have been added to the configuration.

 The preceding example does not show a complete configuration; all regions must have links and routes.

There's more...

As with the other tasks, this task can also be completed using PowerShell.

The following are the three steps to add region links using PowerShell:

1. Create a CSV file with all the information required, which looks similar to this:

2. Save the file (the example shows `C:\LyncScripts\subnets.csv`).
3. Run the following PowerShell command to import and configure:

   ```
   Import-Csv C:\LyncScripts\regionlinks.csv | foreach {
   New-CsNetworkRegionLink -Identity $_.Identity`
    -NetworkRegionID1 $_.NetworkRegionID1`
    -NetworkRegionID2 $_.NetworkRegionID2`
    -BWPolicyProfileID $_.BWPolicyProfileID
   }
   ```

Proceed with the following steps to add region routes using PowerShell:

1. Create a CSV file with all the information required, which looks similar to this:

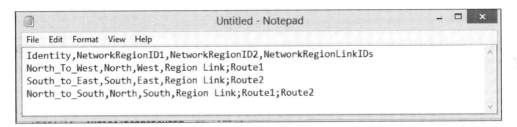

 The `NetworkRegionLinkIDs` field will be populated with two or more existing links. Make sure they are separated by `;` in the CSV file. The `-replace` command will take care of changing the `;` into `,` before executing the commands.

2. Save the file (the example shows `"C:\LyncScripts\RegionRoutes.csv`).
3. Run the following PowerShell command to import and configure:

   ```
   Import-Csv C:\LyncScripts\RegionRoutes.csv| foreach {
   $_.NetworkRegionLinkIDs = $_.NetworkRegionLinkIDs`
    -replace ';',','
   New-CsNetworkInterRegionRoute -Identity $_.Identity`
    -NetworkRegionID1 $_.NetworkRegionID1`
    -NetworkRegionID2 $_.NetworkRegionID2`
    -NetworkRegionLinkIDs $_.NetworkRegionLinkIDs
   }
   ```

4. It is also possible to connect two sites that are directly connected through WAN but do not belong to any region. For this, we must use the **New-CsNetworkInterSitePolicy** command (it is not available through Lync CSCP). Simply run the following in an elevated administrative prompt (the Medium bandwidth policy is just an example):

```
New-CsNetworkInterSitePolicy`
 -InterNetworkSitePolicyID Southwest_Southeast`
 -NetworkSiteID1 Southwest`
 -NetworkSiteID2 southeast`
 -BWPolicyProfileID Medium
```

See also

- The *New-CsNetworkRegionLink* post at http://technet.microsoft.com/en-us/library/gg398437.aspx
- The *New-CsNetworkInterRegionRoute* post at http://technet.microsoft.com/en-us/library/gg398779.aspx
- The *New-CsNetworkInterSitePolicy* post at http://technet.microsoft.com/en-us/library/gg412844.aspx

Enabling CAC

After adding all the relevant network information in the previous steps, it is now time to enable CAC.

Getting ready

Enabling CAC must be done for the sites in Topology Builder and globally in CSCP or PowerShell.

How to do it...

The following are the six steps to enable CAC in sites:

1. Start Topology Builder with an account that has the **RTCUniversalServerAdmins** rights and save the topology file when prompted.
2. For each site in the topology:
   1. Right-click on the site node.
   2. Select **Edit properties**.

3. Scroll down to the **Call admission control setting** option.
4. Select the **Enable call admission control** checkbox, as shown in the following screenshot:

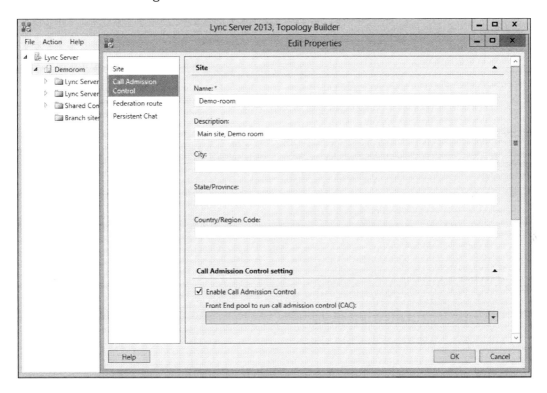

5. If there is more than one pool in the site, select a pool (in a single pool site, the pool should be selected automatically).

 It is common to add this feature to the first pool in each site, but it does not have to be. After we select a pool for this job, the CAC service has to be installed on the selected pool.

6. Click on **OK**.

3. Publish the topology (and make note of the to-do list).
4. Initiate the replication (`Invoke-CsManagementStoreReplication`).
5. Wait for the replication to complete.
6. Complete the to-do list from the publishing wizard.

Proceed with the following steps to enable CAC globally:

1. Launch the Lync Server Control Panel with a **CSAdministrator** account, and navigate to the **Global** option under the **Network Configuration** section.
2. Simply select **Enable call admission control**, as shown in the following screenshot:

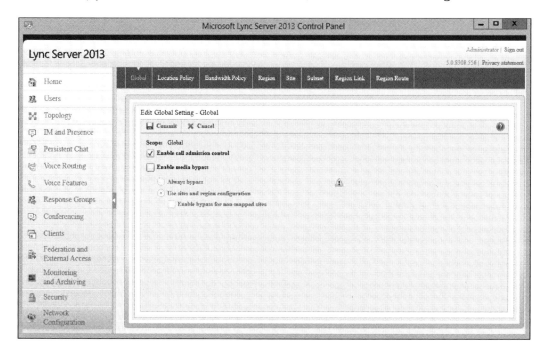

3. Select **Commit**.

 We'll discuss Media Bypass in a different section.

There's more...

There are a lot of site settings that we can manage through the `Set-CsSite` command, but **Enable call admission control** is not one of them. This task has to be completed in Topology Builder.

Chapter 11

The network configuration portion, however, can easily be performed through PowerShell in the following two steps:

1. Launch PowerShell in an elevated prompt.
2. Type the following command:

   `Set-CsNetworkConfiguration -EnableBandwidthPolicyCheck 1`

> As an addition to enabling CAC, it might be a good idea to enable logging as well. This can be done with the following command:
>
> `Set-CsBandwidthPolicyServiceConfiguration -EnableLogging $true - LogCleanUpInterval 10.00:00:00 -MaxLogFileSizeMb 10 -MaxTokenLifetime 12:00:00`
>
> Consider how much data to keep, and adjust the parameters of this command as needed.

The previous command will enable logging and store data in the fileshare under the application server's `\PDP` directory, for example, `\\fileserver.mydomain.com\Share\1-ApplicationServer-1\AppServerFiles\PDP`. These files can be viewed in the Lync Resource Tool Kit **Bandwidth Utilization Analyzer** for further analysis.

See also

- The *Set-CsNetworkConfiguration* post at `http://technet.microsoft.com/en-us/library/gg398642.aspx`
- The *Set-CsBandwidthPolicyServiceConfiguration* post at `http://technet.microsoft.com/en-us/library/gg412863.aspx`

Preparing servers and clients for DSCP tagging

The only way for a network infrastructure to recognize and prioritize the different data flows is if there is something in the data to identify the type of traffic. This can be done in many ways, some more intelligent than others. However, one sure way to alert the infrastructure is to tag the packets with predefined DSCP marking (which must be agreed upon and honored end-to-end in the network).

Getting ready

This task cannot be completed without proper planning and an agreement with the network team on which values to mark the different data types with. We also need to know which ports the different types of traffic will use. Once this is settled, it's time to configure servers and clients to tag the traffic.

 A good starting point could be these example values: Audio DSCP 46 (EF), Video DSCP 34 (AF41), Signaling 24 (CS3), and file transfer 18.

How to do it...

There are at least three systems that we need to address: clients, domain servers, and non-domain servers (edge). The first two can have group policies assigned to them, but the third system must be edited through the local group policy.

The following are the steps to enable QoS system-wide:

1. Launch PowerShell in an elevated administrative mode.
2. Type the following command:

   ```
   Set-CsMediaConfiguration -EnableQoS $true
   ```

Follow these steps to add Client Group Policies. Each type of traffic will need its own QoS policy. Here is the description of the first policy for the clients. Repeat the procedure for all the types of traffic to be tagged as follows:

1. Log on to a server/client with administrative rights to the domain and with the **group policy management tool** available.
2. Start the **Group Policy Management** window.
3. Locate the container/OU where the new policy should be created.
4. Right-click on the appropriate container and then select **Create a GPO in this domain, and Link it here**.
5. Give the new policy a name and then click on OK.
6. Right-click on the newly-created policy and then click on **Edit**.
7. Expand **Computer Configuration** and **Windows Settings**.
8. Right-click on **Policy-Based QoS** and click on **Create new policy**.

9. Type a descriptive name for the new policy. Give it a descriptive name for later reference, for example, `Lync Audio QoS`.
10. Select the **Specify DSCP Value** checkbox, and then select the assigned DSCP (46 for audio).
11. Uncheck the **Specify Outbound Throttle Rate** checkbox.
12. Click on **Next**.
13. Click on **Only applications with this executable name**.
14. Type `communicator.exe` (the Lync client name).
15. Click on **Next**.
16. Leave the source and destination as is (the default is **all source/destination IP**).
17. Click on **Next**.
18. Select **TCP and UDP**.
19. Specify the source port number by selecting **From this source port number or range**. Enter the port range for the media type that you specified while planning.
20. Click on **Finish**.
21. Repeat steps 8 to 20 for all of the media types to tag.
22. Save the GPO, and make sure that the Group Policy is applied to all of the clients where Lync is used.

> To verify whether a Group Policy has been applied to the computer, open a command prompt in the elevated administrative mode and type `gpresult /r /scope COMPUTER`. Look for the policy under **Applied Group Policy Objects**.

Lync Servers (not Edges)

Follow the same procedure to add policies to clients when applying to servers, with the following exception.

> The port settings are applied to the services in the next section, so these tasks must be completed at the same time (more or less) to be of any help.

All server policies must be configured with the **All programs** and **Source ports** settings, and not with **Only applications with this executable name** as the client-side policies are.

Controlling Your Network – A Quick Drill into QoS and CAC

The Edge Server internal interface

Follow the same procedure to add policies to clients when applying to servers, with the following exceptions:

- The Edge Server cannot receive Group Policies from the domain. The configuration must be performed on the server in **Local security Policy**.
- All edge server policies must be configured with the **All programs** and **Destination ports** settings, and not with **Only applications with this executable name** as the client-side policies are.

Controlling/limiting the port ranges for traffic

So far, we've discussed how to control the number of sessions allowed between sites (CAC) and how to mark packets based on ports and applications. However, we have not yet discussed how to tell the servers and clients which port ranges it should use for the different types of media. Without doing this, Lync will use the default ports, and our marking and QoS scheme might fail.

> Values given here are examples only! Adjustments might be necessary for larger or smaller deployments. The number of available ports can influence the call flows available in a system. Also, remember that the port number assignments must never overlap (how will QoS tagging know the difference between services?).

Getting ready

There are a lot of components in Lync, and we need to address all of those services. Luckily, all the settings are stored in the topology, and we can apply all of the settings through PowerShell from the same management station.

We will configure conferencing services, mediation services, application services, and edge services.

How to do it...

Proceed with the following steps for controlling/limiting the port ranges for traffic:

1. Log on to a server with Lync administrative tools, and launch PowerShell in the elevated administrative mode.

    In most deployments with collocated services, the application pool, mediation pool, and conferencing pool will have the same pool name.

2. Enable QoS system-wide by typing the following command:

   ```
   Set-CsMediaConfiguration -EnableQoS $True
   ```

3. Enable QoS for the conferencing server. For each conferencing pool in the system, type the following command:

   ```
   Set-CsConferenceServer -Identity "myconfpool.mydomain.com" `
    -AudioPortStart "50000" -AudioPortCount "2000" `
    -VideoPortStart "52000" -VideoPortCount "2000" `
    -AppSharingPortStart "54000" -AppSharingPortCount "2000"
   ```

4. Enable QoS for the mediation server. For each mediation pool in the system, type the following command:

   ```
   Set-CsMediationServer -Identity "mymedpool.mydomain.com" `
    -AudioPortStart "50000" -AudioPortCount "2000"
   ```

    Don't configure AudioPortCount, which is too low. A single audio call will need at least 7 calls.

5. Enable QoS for the application servers. For each application pool in the system, type the following command:

   ```
   Set-CsApplicationServer -Identity "myapppool.mydomain.com" `
    -AudioPortStart "50000" -AudioPortCount "2000" `
    -VideoPortStart "52000" -VideoPortCount "2000" `
    -AppSharingPortStart "54000" -AppSharingPortCount "2000"
   ```

6. Enable QoS for the Edge Servers. For each Edge pool in the system, type the following command:

   ```
   Set-CsEdgeServer -Identity "myedgepool.mydomain.com"`
     -MediaCommunicationPortStart "50000"`
     -MediaCommunicationPortCount "10000"
   ```

7. Enable QoS for the clients. Type the following command:

   ```
   Set-CsConferencingConfiguration`
     -ClientMediaPortRangeEnabled $True`
     -ClientAudioPort "50000" -ClientAudioPortRange "2000"`
     -ClientVideoPort "52000" -ClientVideoPortRange "2000"`
     -ClientAppSharingPort "54000"`
     -ClientAppSharingPortRange "2000" `
     -ClientFileTransferPort "56000"*
     -ClientFileTransferPortRange "2000"
   ```

See also

- The *Set-CsMediaConfiguration* post at http://technet.microsoft.com/en-us/library/gg398580.aspx
- The *Set-CsConferenceServer* post at http://technet.microsoft.com/en-us/library/gg398738.aspx
- The *Set-CsMediationServer* post at http://technet.microsoft.com/en-us/library/gg398213.aspx
- The *Set-CsApplicationServer* post at http://technet.microsoft.com/en-us/library/gg398562.aspx
- The *Set-CsEdgeServer* post at http://technet.microsoft.com/en-us/library/gg398859.aspx
- The *Set-CsConferencingConfiguration* post at http://technet.microsoft.com/en-us/library/gg412969.aspx

Media bypass

The function of media bypass is to allow endpoints to send media traffic (audio) directly to the PSTN gateway, and use the mediation pool only for signaling. This greatly improves the number of calls that a mediation server can handle. When media bypass is not selected, each co-located mediation server can handle up to 155 simultaneous calls (a dedicated mediation server can handle up to 1,500 calls). With media bypass enabled, the performance is greatly improved.

Media bypass and CAC do not go hand in hand, and this is why the **always bypass** option is disabled when **Enable CAC** is selected. It is said that CAC and media bypass are mutually exclusive. There are scenarios where we might possibly not need CAC in a deployment. It is worth going through the options and looking at a few scenarios.

When would it be possible to use media bypass in a deployment where QoS and CAC are implemented?

In a large deployment with several sites and PSTN trunks, or simply in a huge campus with a PSTN trunk, it could be of real benefit to have those clients with the same bypass ID as the trunk to bypass the mediation server. This will offload the mediation servers and keep resources available for those calls that come across the WAN, which need CAC to be maintained.

Whenever QoS and CAC are implemented together, we need to use the **Use Site and Region Information** option where there has to be a match between the caller and the called party's bypass ID.

Getting ready

To complete this task, we need CSCP and an understanding of sites/links where media bypass can be utilized.

How to do it...

1. Launch the Lync Server Control Panel with a **CSAdministrator** account, and navigate to **Network Configuration | Global**.
2. CAC should have already been selected, but now is the time to add **Enable media bypass**.
3. Select **Use Site and Region Information** and **Enable bypass for non-mapped sites**. The latter will enable sites that are not identified (not configured) to use media bypass.
4. Next, navigate to the **trunk configuration** option in the **Voice routing** section.
5. Select the trunk where bypass will be used (site trunks).
6. Select **Enable media bypass** on the trunk.
7. Select **Commit**.

12
Lync 2013 Debugging

In this chapter, we will cover the following topics:

- Introduction
- Using Snooper to examine log files
- Investigating Call Flow with Snooper Flow Chart
- Reviewing Lync information with OCSLogger
- Tracing from a command line with OCSTracer
- Customizing CLS scenarios using CLSController
- Testing our setting with Best Practices Analyzer
- Capturing network traffic with Wireshark
- Troubleshooting clients with the Microsoft Lync Connectivity Analyzer
- Verifying a deployment with the Microsoft Remote Connectivity Analyzer

Introduction

Troubleshooting Lync Server 2013 is a multifaceted task, whose complexity is related to the kind of deployment we are working on. For instance, it is much quicker and easier to troubleshoot issues with a single site Standard Edition Pool without Enterprise Voice than it is for a larger deployment with multiple pools and more features. This chapter is dedicated to some practical examples and hints related to the fundamental tools that we have at our disposal to debug Lync. This chapter also contains references to external resources that can be helpful in understanding the potential of the previously mentioned tools and in designing actual troubleshooting procedures.

The tools we will talk about will cover a large part of the different areas involved in Lync debugging, such as protocol stack analysis and log parsing. *Chapter 11, Controlling Your Network – A Quick Drill into QoS and CAC*, also contains information about QoE and network data quality inspection.

Using Snooper to examine log files

Snooper is a tool used to analyze log files in Lync Server. It is a tool we will use over and over to read different kinds of logs and information that come from Lync Servers and endpoints. While Snooper has been available since **Microsoft Lync Server 2010 Resource Kit**, some of the ways we use it have changed with Lync Server 2013 because the server architecture now includes an element called **Centralized Logging Service** (**CLS**). CLS is a great alternative for the previously used solution, the **OCSLogger**, which required a lot of work to gather data from multiple servers. The architecture is based on a service executable, the **Centralized Logging Service Agent** (`ClsAgent.exe`), which runs on each Lync 2013 server, and waits for commands from the service controller `CLSControllerLib.dll` that sends Start, Stop, Flush, and Search commands to the ClsAgent. The ClsAgent uses scenarios that define which Lync components (providers) at what level will be captured in the traces. There are default scenarios and custom scenarios (created with the `New-CsClsScenario` command).
Two scenarios can be run on a given computer at any one point in time. This means you can run a default or a custom scenario in addition to **AlwaysOn** logging (a special scenario that can be enabled to log all the time). We will see how to enable CLS and how to use Snooper to examine the logs it generates.

Getting ready

Snooper is part of the **Lync Server 2013 Debugging Tools**, and we are able to download them from the Microsoft site at `http://www.microsoft.com/en-us/download/details.aspx?id=35453`, or we can use the `Set-Cs2013Features.ps1` script from Pat Richard's site (`http://www.ehloworld.com/1697`). I always suggest that you use the previously mentioned script to automate the installation of the prerequirements for Lync Server roles and for an easy download and installation of additional tools.

How to do it...

1. To see a complete list of the existing scenarios, we can use the `Get-CsClsScenario` cmdlet. A server with the latest updates applied (right now, Microsoft Lync Server 2013 and Core Components 5.0.8308.803) will show 29 scenarios.

2. To view the details about a single scenario, we can use a cmdlet like the following one (for the media connectivity scenario):

   ```
   Get-CsClsScenario global/mediaconnectivity | Select
   -ExpandProperty Provider | FT Name,Level,Flags -Auto
   ```

   The result will be the one shown in the following screenshot (including the flags for monitored information and the log level):

   ```
   PS C:\Users\administrator.WONDERLAND> Get-CsClsScenario global
   | Select -ExpandProperty Provider | FT Name,Level,Flags -Auto

   Name                              Level  Flags
   ----                              -----  -----
   MediaStack_AUDIO_AGC              Info   All
   MediaStack_AUDIO_DRC              Info   All
   MediaStack_AUDIO_ECHODT           Info   All
   MediaStack_AUDIO_FAXDT            Info   All
   MediaStack_AUDIO_HEALER           Info   All
   MediaStack_AUDIO_NOISEDT          Info   All
   MediaStack_AUDIO_VAD              Info   All
   MediaStack_AUDIO_VSP              Info   All
   MediaStack_AudioCodecs            Info   All
   MediaStack_AudioEngine            Info   All
   MediaStack_COMAPI                 Info   All
   MediaStack_COMMON                 Info   All
   MediaStack_Crossbar               Info   All
   MediaStack_Crypto                 Info   All
   MediaStack_DebugUI                Info   All
   MediaStack_DebugUI_AEC            Info   All
   MediaStack_DEVICE                 Info   All
   MediaStack_MassConvertedTraces1   Info   All
   MediaStack_MediaManager           Info   All
   MediaStack_PerFrame               Info   All
   MediaStack_PerPacket              Info   All
   MediaStack_QualityController      Info   All
   ```

3. After we have selected the scenario that best fits our needs, we have to launch the CLS process with the following cmdlet (note that **madhatter.wonderland.lab** is the name of the Front End pool):

   ```
   Start-CsClsLogging -Scenario mediaconnectivity -Pools madhatter.
   wonderland.lab
   ```

   > If you have an Edge pool/server, you might decide not to use the -Pools parameter, so logging will be enabled on all servers.

We will receive a confirmation message, like the one in the following screenshot:

```
PS C:\Users\administrator.WONDERLAND> Start-CsClsLogging -Scenario
vity -Pools madhatter.wonderland.lab
Success Code - 0. Successful on 1 agents


Tracing Status:

madhatter.wonderland.lab (madhatter v5.0.8308.0) (AlwaysOn=No,Scen
ectivity,Started=9/5/2014 2:40:51 PM,By=WONDERLAND\administrator,D
```

4. Use the `Stop-CsClsLogging` command (otherwise, logging will continue to capture data for four hours, which is the default). Another option is to use the `Sync-CsClsLogging` command (in our scenario, **Sync-CsClslogging –Pools madhatter.wonderland.lab**). Both the `Stop-CsClsLogging` and `Sync-CsClsLogging` commands will write the information stored in memory to an ETL file that we are able to search.

5. To make use of the results generated from CLS, we must use a cmdlet like the following:

   `Search-CsClsLogging -OutputFilePath "C:\CLSresult.log"`

6. It is also possible to insert only information from specific components in the log, with a cmdlet like the following one, focused on RTCP:

   `Search-CsClsLogging -Components MediaStack_RTCP -OutputFilePath "C:\Transport.log"`

7. Now, it is useful to launch `Snooper.exe` from the Lync debugging tools installation folder (for example, `C:\Program Files\Microsoft Lync Server 2013\Debugging Tools`).

8. Select **File | Open file** and click on one of the previously generated logs.

Chapter 12

9. Snooper will parse the log and create two views: one dedicated to the visualization of all the information available (**Trace**), and the second with a list of messages organized in groups, with the errors on the left (**Messages**), similar to what we are able to see in the following screenshot:

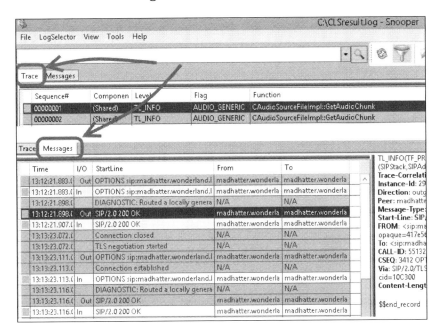

 To install the Lync 2013 Debugging Tools on a workstation, the prerequirements include .NET Framework, PowerShell 3.0, and the Visual C++ 2012 Runtime. Note that the Visual C++ version required is 11.0.50727 (the one on the Lync Server 2013 ISO image).

There's more...

You may also rely on third-party tools to ease the Centralized Logging Server management. We describe one of these in *Chapter 5, Scripts and Tools for Lync*, in the *Tracing made easier – Lync 2013 Centralized Logging Tool* recipe.

Lync 2013 Debugging

See also

- There is an interesting session from the TechEd North America 2014 titled *Advanced Troubleshooting for Microsoft Lync* at `http://channel9.msdn.com/Events/TechEd/NorthAmerica/2014/OFC-B411#fbid=`, which is focused on troubleshooting and the related tools in Lync, including Snooper. The speakers propose a systematic approach to troubleshooting, which could be important to keep in mind.

Investigating Call Flow with Snooper Flow Chart

The **Flow Chart** (**Call Flow**) is a visualization mode in Snooper that shows a diagram of an SIP-based communication or call. Although it does not add information to what we already see in the messages, this kind of outline is helpful in examining the various steps of the call in a single view. We will see a couple of examples in this section.

How to do it...

1. Launch Snooper, open a logfile, and go to messages, as we have seen in the previous section.

2. Select a message on the left and then click on the **Show Call Flow window** option in the top-right menu, as shown in the following screenshot:

Chapter 12

3. In the first example, we have a conversation between a Lync client in the internal network and a Skype user (federation), as shown in the following screenshot:

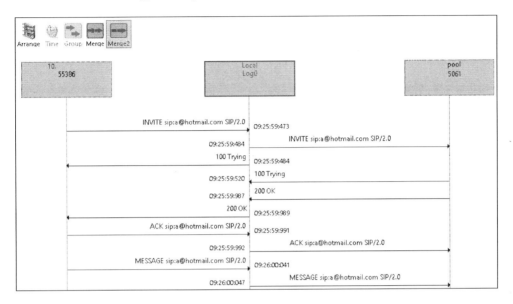

4. The second example is a call from a Lync client to a mobile phone, as shown in the following screenshot:

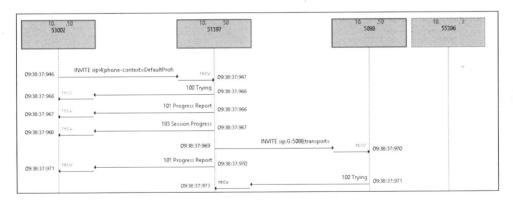

There's more...

In the post *Lync Media Conferencing/Audio call flow of Sip and SDP* at `http://digitalbamboo.wordpress.com/2013/08/17/lync-media-conferencingaudio-call-flow-of-sip-and-sdp/`, we take an interesting deep dive on the call flow in Lync.

Reviewing Lync information with OCSLogger

The **Centralized Logging Service** (**CLS**) that we mentioned is for sure a powerful feature to troubleshoot Lync issues. However, in some specific scenarios, for example, when we need to quickly change the protocols that we are logging, we prefer to use a different tool to trace the protocol stack, the `OCSLogger.exe` tool.

Getting ready

Like Snooper, it is part of the Lync Server 2013 Debugging Tools. There are a couple of advantages to using OCSLogger over CLS. First and foremost, we should know that we have a graphical interface, rather than PowerShell, to manage the logging parameters.
The second one is that we are able to change the protocols and features that we are tracing in a fast-paced manner. We will take a quick look at how the OCSLogger works.

How to do it...

1. If we have any CLS scenarios activated, the OCSLogger will not run. The error that we encounter is **OcsLogger Application may not be run when Centralized Logging Service is Running**. Our first step must be to verify whether a tracing is running using the `Show-CsClsLogging` cmdlet. In the following screenshot, we have the AlwaysOn scenario active on our Lync pool:

```
PS C:\Users\administrator.WONDERLAND> Show-CsClsLogging
Success Code - 0, Successful on 1 agents


Tracing Status:

madhatter.wonderland.lab (madhatter v5.0.8308.0) (AlwaysOn=Yes)
    madhatter.wonderland.lab (madhatter v5.0.8308.0) (Same as pool)

PS C:\Users\administrator.WONDERLAND>
```

Chapter 12

2. Our example pool is **madhatter.wonderland.lab**. The cmdlet to be used to stop the previously mentioned trace is as follows:

   ```
   Stop-CsClsLogging -Scenario "AlwaysOn" -Pools madhatter.wonderland.lab
   ```

   The tracing status will change to (**AlwaysOn=No**).

3. It is important to add a couple of notes here:

   - It might be necessary to launch the `Stop-CsClsLogging` command more than once. Sometimes, we have an onscreen error, or we see that 0 agents are affected by the command. Launching it more than once usually resolves the issue.
   - It is important to use the quotation marks around the scenario name.

4. Now, we are permitted to launch the `OCSLogger.exe` tool from its installation path (for example, `C:\Program Files\Microsoft Lync Server 2013\Debugging Tools`).

5. In the following two screenshots, we can see a quick recap of the options available on the main screen. The following screenshot is for the left side of the tool:

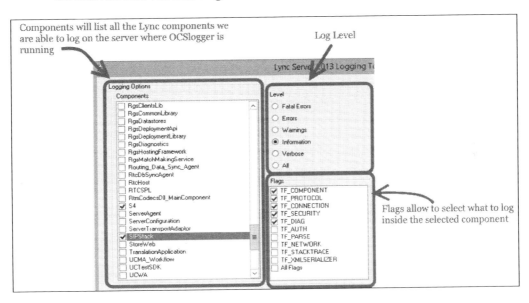

343

The following screenshot shows the options on the right side:

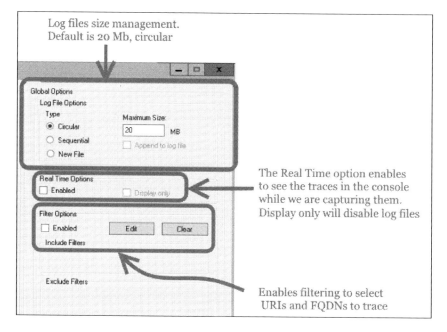

6. In OCSLogger, the default choices for components are **Collaboration**, **S4**, **SpeechComponent**, and **SpeechVxmlComponent**. SIPStack and S4 are the ones we will use more often to debug issues with the SIP protocol.

7. For a complete list of all the options available inside the OCSLogger, there is a handy post on TechNet, which specifies the options for logging at `http://technet.microsoft.com/en-us/library/bb936621(v=office.12).aspx`.

8. The **Advanced Options** window for the OCSLogger includes a **Formatting** tab (to customize the way information is stored in the logs), **Buffering** (to modify the default buffer values for real-time monitoring), **Clock Resolution** (to define the timestamps in the log), and **Additional Components** (to enable logging of more Lync components). A quick overview of the options is shown in the following screenshot:

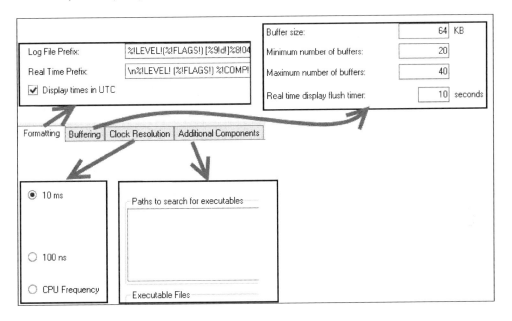

9. Details about the advanced options are available in the *Using Lync Server 2013 Logging Tool* post at http://msdn.microsoft.com/en-us/library/office/dn466134(v=office.15).aspx.

10. When we have set all the options to fit our requirements, we can begin tracing using the **Start Logging** button. It is not required to keep the interface opened. Closing the OCSLogger GUI will prompt us if we want to actually stop the logging process.
If we select **No**, the tracing will continue until we stop it.

11. By selecting the **Stop Logging** button, we will interrupt the tracing operations. It is possible to view logfiles or to analyze logfiles. We will receive a request to select the components to analyze.

12. Snooper will be launched on screen, with our logs opened.

Lync 2013 Debugging

There's more...

As we have seen, the OCSLogger will keep tracing information until we stop it (if we simply close the interface without stopping the log registration, logs will still be generated).
If we are not using circular logging, it is important to watch the available disk space because files might increase in size really fast.

See also

- Some time ago, Jeff Schertz published an interesting post about *Using the Lync Logging Tool* at `http://blog.schertz.name/2011/06/using-the-lync-logging-tool/`. It contains some interesting points, which are still worthy of reading.

Tracing from a command line with OCSTracer

The `OCSTracer.exe` tool is a command-line utility (part of the Lync Server 2013 Debugging Tools, as the previous tools), which enables logging from a command line.
We will see some practical examples in this section.

How to do it...

1. From a command line, go to the `C:\Program Files\Microsoft Lync Server 2013\Debugging Tools` folder and run the following command to log the S4 component, with verbose level (TL_Verbose) on the TF_Component as follows:

   `ocstracer start /component:S4,TL_Verbose,TF_Component`

2. To stop all the tracing, we can use the `ocstracer stop` command. If we want to stop only a part of the tracing, we have to launch a command similar to the one used for the launch, replacing `start` with `stop`.

3. We will receive an on-screen notification about the number of buffers and events recorded, as shown in the following screenshot:

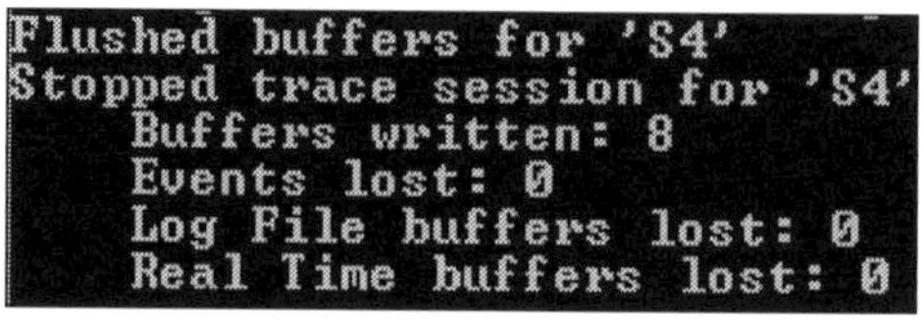

```
Flushed buffers for 'S4'
Stopped trace session for 'S4'
    Buffers written: 8
    Events lost: 0
    Log File buffers lost: 0
    Real Time buffers lost: 0
```

Chapter 12

4. To select a path and a filename for the log, we can use the following command (the selected path for the example is `c:\temp`, and the filename is `test.txt`):

   ```
   Ocstracer start /component:S4,TL_Verbose,TF_Component /
   logfilefolder:c:\temp /logfilename:test.txt
   ```

To modify the default settings for OCSTracer, it is also possible to edit the `OCSTracer.ini` file. All the default starting parameters, including the log file's path, are stored here.

Customizing CLS scenarios using CLSController

Another tool included in the Lync Server 2013 Debugging Tools is the `ClsController.psm1` file. It is a PowerShell module used to customize the CLS scenarios and can be tailored to satisfy our troubleshooting needs. In the module, we have a cmdlet, `Edit-CsClsScenario`, which we will use for an example of customization.

How to do it...

1. Launch the Lync Management Shell and use the following cmdlet:

   ```
   Import-Module "C:\Program Files\Microsoft Lync Server 2013\
   Debugging Tools\ClsController.psm1"
   ```

2. We can check whether the module has been loaded correctly using the following command:

   ```
   Get-Command -Module ClsController
   ```

3. It is advisable to define a new scenario. The first step is to define the provider:

   ```
   $testprov = New-CsClsProvider -Name "Testprov" -Type "WPP" -Level
   "Info" -Flags "All"
   ```

4. Then, we have to define a new policy:

   ```
   New-CsClsScenario -Identity "global/TestprovScen"
   -Provider $testprov
   ```

5. We will receive a confirmation message, as shown in the following screenshot:

   ```
   Identity : Global/TestprovScen
   Provider : {Name=Testprov;Type=WPP;Level=Info;Flags=All;Guid=;Role=}
   Name     : TestprovScen
   ```

347

Lync 2013 Debugging

> If we are going to customize an existing scenario, we need to know the providers we want to remove or add. To see a list of the default providers, we can use the `Get-CsClsScenario` cmdlet. For example, for `TestprovScen`, use the following:
>
> `$scenario=Get-CsClsScenario global/TestprovScen`
>
> `foreach ($sc in $scenario.provider) { $sc.name }`

6. To add a provider, like the ServerAgent (with verbose level and all the flags selected), we can use the following command:

   `Edit-CsClsScenario -ScenarioName TestprovScen -ProviderName ServerAgent -Level Verbose -Flags All`

7. By running `Get-CsClsScenario`, as we have seen before, we will get the result as shown in the following screenshot. The **ServerAgent** has been added to the scenario.

   ```
   PS C:\U> Edit-CsClsScenario -ScenarioName TestprovScen -ProviderName ServerAgent -Level Verbose -Flags All
   PS C:\U> $scenario=Get-CsClsScenario global/TestprovScen
   PS C:\U> foreach ($sc in $scenario.provider) { $sc.name }
   Testprov
   ServerAgent
   ```

8. To remove a provider, add the `-remove` option. There is no support for the `-Level` and the `-Flags` option, so to remove the ServerAgent provider in the previously mentioned example, we should use the following:

   `Edit-CsClsScenario -ScenarioName TestprovScen -ProviderName ServerAgent -Remove`

See also

- For the cmdlet in step 3, I have used a syntax that is similar to the one in this *Enumerating Lync 2013 Central Logging Scenarios* post at http://blogs.catapultsystems.com/IT/archive/2012/12/20/enumerating-lync-2013-central-logging-scenarios.aspx. I suggest that you read the full post, which is very useful when manipulating scenarios.
- The TechNet blog deep dives on some aspects of CLS in the *Lync 2013 Centralized Logging (CLS) - customizing scenarios* post at http://blogs.technet.com/b/rischwen/archive/2014/02/10/lync-2013-centralized-logging-cls-customizing-scenarios.aspx.

Chapter 12

Testing our setting with Best Practices Analyzer

Lync Server 2013 has a dedicated software, the Lync Server 2013 **Best Practices Analyzer** (**BPA**), which is used to perform a series of configuration checks on our deployment. BPA is usually a proactive tool used to identify wrong settings, missing updates, and suggested configurations that we are missing. Let's take a quick look at it.

Getting ready

The BPA is available for download at `http://www.microsoft.com/en-us/download/details.aspx?id=35455`. Its setup is really easy, although it is important to ensure that an Internet connection is available to update the information that BPA uses to analyze the system.

How to do it...

1. After we have downloaded and installed the BPA, we will be able to launch it from the installation folder (the default path is `C:\Program Files\Microsoft Lync Server 2013\BPA\RTCBPA.exe`) or from its icon on the start screen.
2. BPA will start with an update from the Internet to align with the latest available version of the best practices.
3. The next step is to open the **Welcome** page. We are able to open an existing scan or perform a new one. We will start a new one.
4. BPA will require a connection to the Active Directory Services Server (`Alice.Wonderland.Lab`, in our scenario) as shown in the following screenshot:

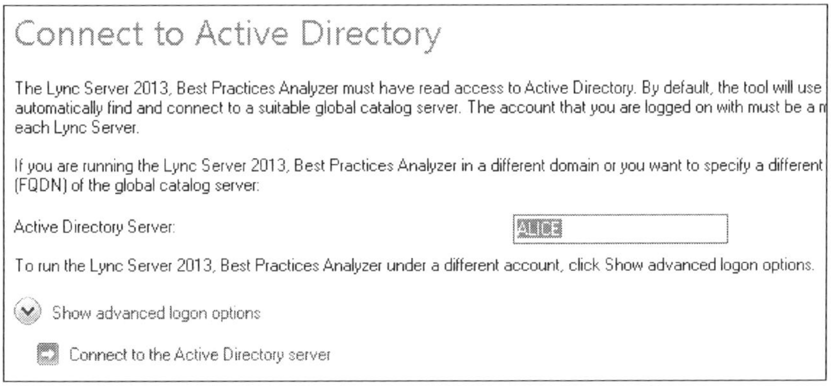

349

5. Select **Connect to the Active Directory Server**.
6. BPA will offer a list of Lync Servers that we are able to check, as we can see in the following screenshot:

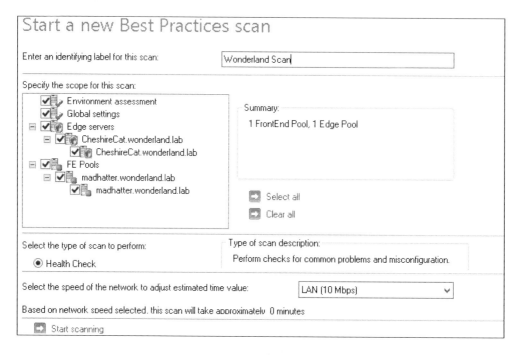

7. At the end of the scan, we have an overview of the outcome of the BPA controls and a link to view a report of this best practices scan.

Chapter 12

8. The report is available with different outlines. Each outline has its own way to show the information. It is possible to sort the results in the different reports based on the class, issue, and severity. For example, we have a **List Reports** selection, as shown in the following screenshot:

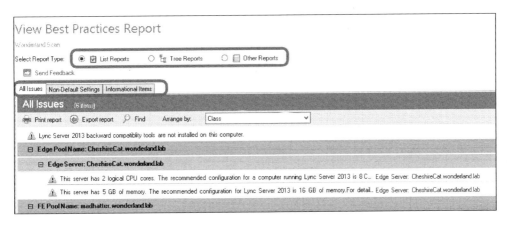

9. Results are made up of information items, warnings, and errors. Some of the results also contain links and recommendations on how to resolve an issue.

Capturing network traffic with Wireshark

Wireshark is an open source (GNU) tool used to capture and analyze network traffic. It is often used to troubleshoot Lync issues or to deep dive into the network traffic related to a specific feature. In addition to this, Wireshark adds to the standard debugging tools the capability to decrypt SSL/TLS traffic. It is really important for issues related to the Lync web services. We will see some hints related to installing and using it.

Getting ready

Wireshark is available for download at `https://www.wireshark.org/`. We will use the latest stable release, 1.12.0, on our Lync Front End (installed on Windows Server 2012 R2) by downloading the matching Windows installer (64 bit) `Wireshark-win64-1.12.0.exe`.

How to do it...

1. Select **Next** on the first installation screen and the **I Agree on the License Agreement** screen.
2. We have to select the components to install (the Wireshark 2 preview is not required), and select **Next**, as shown in the following screenshot:

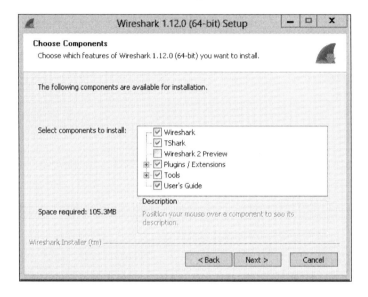

3. In the **Select Additional Tasks** window, we are able to select the launch icons to create and the trace files associated to Wireshark. Click on **Next**, as shown in the following screenshot:

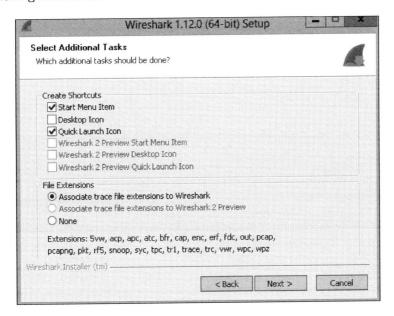

4. Select an installation path and click on **Next**.

5. Install WinPcap, as shown in the following screenshot. Without WinPcap, we cannot capture live network traffic but only open saved capture files.

6. The Wireshark and WinPcap installation will start. WinPcap will require additional confirmation regarding the installation path and startup mode.
7. Launch Wireshark and select the interface for the capture, and then click on **Capture Options**, as shown in the following screenshot:

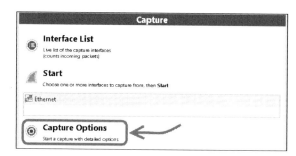

8. An important decision that we have to take is whether we want to capture a large amount of network traffic and then filter it, or whether we prefer to filter traffic directly before displaying it in Wireshark. We have some predefined filters, and we are also able to define our custom selection. The default filters include IP, TCP only, and UDP only. To make use of additional filters, we will need to enter them using a specific syntax. For example, to filter all SIP-related traffic, we can use a string like `tcp port 5041 or tcp portrange 5061-5065 or tcp portrange 5067-5068 or tcp portrange 5071-5073 or tcp portrange 5075-5076 or tcp portrange 5081-5082 or tcp portrange 5086-5087.`

The following screenshot shows the **Wireshark Capture Options** screen:

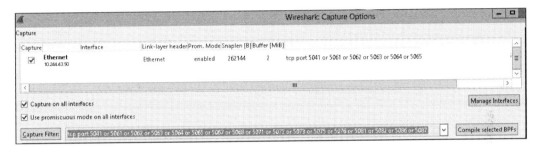

9. It is possible to save custom filters with a name inside Wireshark. We will see two different scenarios.

10. Starting with a broad filter like **IP**, we have to apply a filter to see only the relevant results. The filter might be a simple one, like **dns**, or the network address of one of our servers, with a filter such as **ip.addr == 192.168.1.10**. We can also use the **Expression...** menu to select from many available options and to apply more complex filters, as shown in the following screenshot:

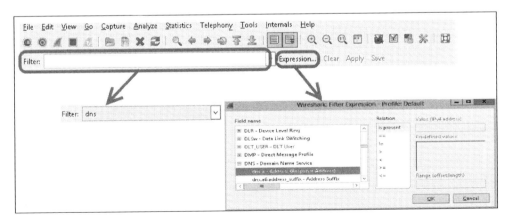

11. Starting with a filter like the custom one that we have in the previous step, Wireshark will only show the network traffic that matches the filter, as shown in the following screenshot:

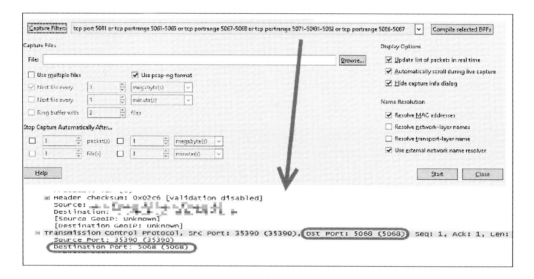

12. If we are troubleshooting an Enterprise Voice issue, Wireshark has a **Telephony** dedicated menu, which includes a **VoIP Calls** option as shown in the following screenshot:

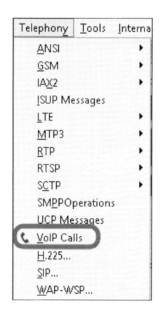

13. The VoIP Calls screen contains two features, **Flow** and **Player**. The **Flow** feature is used to see the call flow as we did with the Snooper Flow Chart. This is shown in the following screenshot:

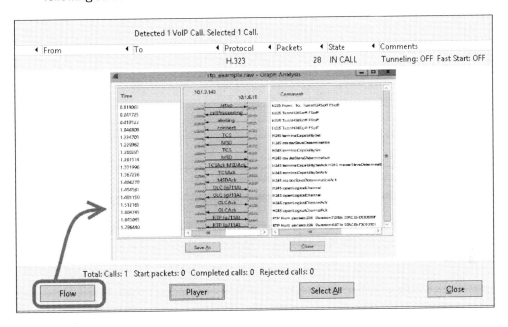

14. The player will open a second window, enabling us to decode the call by using the **Decode** option and then, in a further window, to listen to it, as shown in the following screenshot:

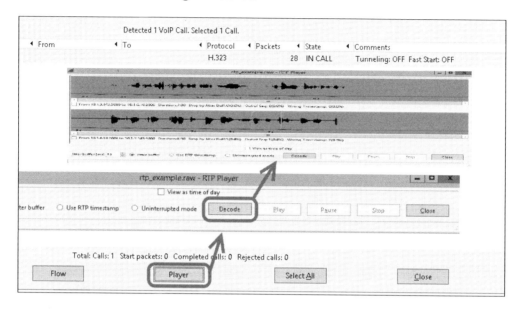

How it works...

To learn more about Wireshark hands-on, there is a useful page at http://wiki.wireshark.org/SampleCaptures, which contains a collection of samples that we can use to get some expertise with the tool.

There's more...

James Cussen has developed an interesting addition for Wireshark, the *Microsoft Lync Wireshark Plugin* at http://www.mylynclab.com/2014/05/microsoft-lync-wireshark-plugin.html. This plugin is meant to help in capturing and decoding audio and video traffic that is entering and exiting from the Lync Edge. Some of the protocols involved are not completely standard and are not readable from Wireshark without some work that the plugin does for us.

Jonathan McKinney has outlined the process to make the Lync-encrypted traffic readable in Wireshark. The main point is to export the SSL certificate from Lync and import it inside the tool. We can read it in the *Using Wireshark to Decrypt Lync Communications* post at http://blog.lyncdialog.com/2013/11/using-wireshark-to-decrypt-lync.html.

See also

- A good introduction to the use of Wireshark with Lync is the one written by Matt Landis, *Getting Started With Lync and Wireshark: Tips & Quirks* at `http://windowspbx.blogspot.com/2013/11/getting-started-with-lync-and-wireshark.html`.

Troubleshooting clients with the Microsoft Lync Connectivity Analyzer

The **Microsoft Lync Connectivity Analyzer** (**MLCA**) is a tool designed to help administrators of a Lync deployment (on-premises or in Office 365) verify whether their architecture is able to support Lync client apps that run on mobile devices. Lync apps require the following specific configurations:

- The DNS records (Lyncdiscover and Lyncdiscoverinternal) should be able to automatically locate Lync services
- A reverse proxy configuration is required to publish the previously mentioned resources
- Specific settings (hairpinning on the reverse proxy) for the clients on the internal network (for example, using a Wi-Fi system) that will connect to the Lync services like a client that comes from an external network

Getting ready

The MLCA is available in two different versions, 32-bit (`http://www.microsoft.com/en-us/download/details.aspx?id=36536`) and 64-bit (`http://www.microsoft.com/en-us/download/details.aspx?id=36535`) ones. The MLCA tool is available on Windows 7, Windows 2008 Service Pack 2, and all the previously mentioned operating systems. It requires the Microsoft .NET Framework 4.5 installed (on Windows Server 2012 R2 and Windows 8, it is a system feature activated by default).

How to do it...

1. We have to download and install the MLCA. The setup process is a standard one, which just requires a confirmation about the installation path (it is advisable to run it with local administrative permissions).

2. MLCA supports four different scenarios to test three different kinds of clients (the Lync Store app, the Lync 2010 app, and the Lync 2013 app). The scenario that we are going to use is defined by the network where the verification will occur (internal or on the Internet) and by the type of deployment we are going to verify (on-premises or in the cloud). All the configuration menus for the tool are shown in this screenshot:

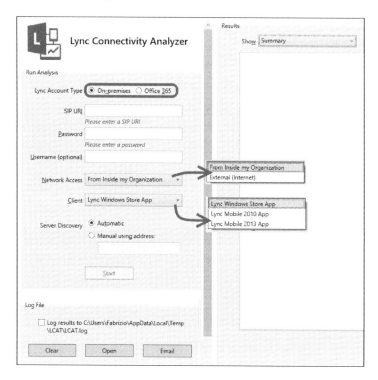

3. Similar to a real Lync app, it is required to insert a user's SIP URI, a password, and a domain account if it does not match with the SIP URI. Discovery can be automatic or manual.

4. For our first test, we can try an Office 365 user that is connecting from the Internet. The results are available in three formats. We are able to see the summary and detailed information in the next screenshot:

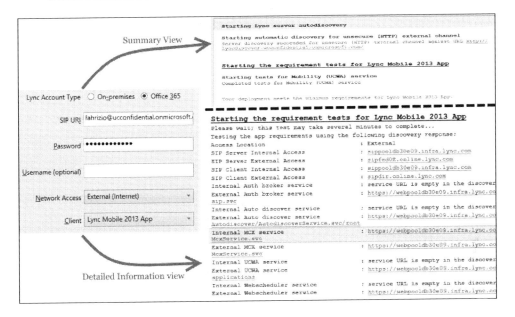

5. The previous test, when configured to simulate an external connection, tried to resolve the autodiscover service using the Lyncdiscover record. As we were analyzing connectivity for an Office 365 user, the Lyncdiscover server was contacted directly using HTTP.

6. If we try to verify an on-premises user connected to the Internet, the test will perform a connection attempt via Lyncdiscover with the HTTPS protocol, and revert to HTTP if a failure occurs.

Lync 2013 Debugging

Verifying a deployment with the Microsoft Remote Connectivity Analyzer

The Microsoft **Remote Connectivity Analyzer** (**RCA**) is a web page that contains a collection of tools to verify the Exchange and Lync functionality, both on-premises and on Office 365.

We have already used the RCA in the *Authenticating with online services using DirSync* recipe of *Chapter 2, Lync 2013 Authentication* to verify our Single Sign-On solution. RCA runs outside our corporate network, so it is one of the best ways we have to check our Lync services from the point of view of an external user. The checks performed by RCA are aligned to the best practices suggested from Microsoft, so we can also use it as a sort of validation for our solutions. We will see the RCA tools dedicated to Lync.

Getting ready

There are no special requirements to use Microsoft RCA other than making the credentials of a Lync-enabled user readily available. We can just go to `https://testconnectivity.microsoft.com/` in an Internet browser and select the test that fits our needs.

How to do it...

1. There are five different tabs available on the RCA page. The **Client** tab only contains links to offline tools used for testing. One of them is the Microsoft Lync Connectivity Analyzer Tool that we saw in the previous section. The **Lync / OCS** page is the one that we are going to use most frequently. The tab is shown in the following screenshot:

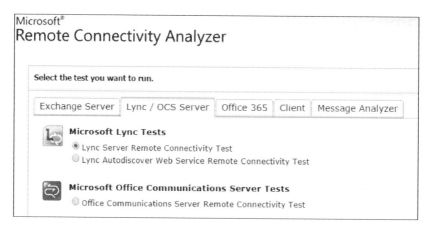

Chapter 12

2. There are two different tests available: **Lync Server Remote Connectivity Test** and **Lync Autodiscover Web Service Remote Connectivity Test**. The Autodiscover test simply requires an SIP URI to check the Lyncdiscover DNS record and the related web service (based on HTTPS), as shown in the following screenshot:

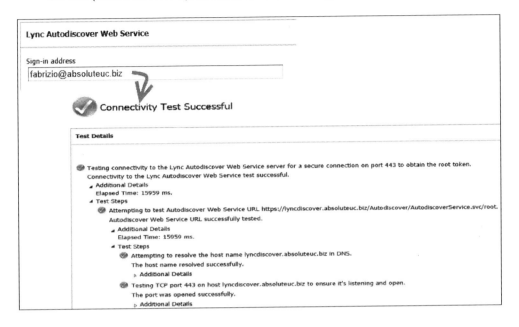

3. The Server Remote Connectivity Test verifies all the steps for an external client that is trying to connect to our Lync deployment. The domain username and password are required in addition to the SIP URI. It is also possible to verify the audio and video functionality. A test similar to the previously mentioned one will verify our Edge deployment. One option enables us to ignore problems with the SSL certificate, and this is a useful option if we are using (for example) certificates from an internal certification authority, as shown in the following screenshot:

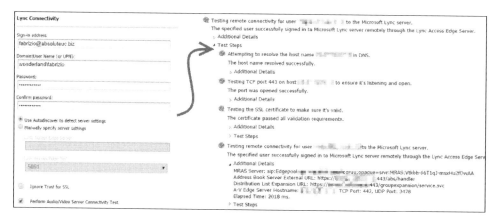

4. If an error occurs, the RCA will show us both the error ID and a connection to TechNet articles and documents to help us resolve the issue.

5. The Office 365 tab includes two tests we could be interested in, **Office 365 Lync Domain Name Server (DNS) Connectivity Test** and **Office 365 Single Sign-On Test**. We have already mentioned the latter. The DNS connectivity test verifies four fundamental records required for our Office 365 subscription to work correctly. They are **_sip._tls**, **_sipfederationtls**, **sip**, and **lyncdiscover**.

Index

Symbols

-ListAsGridView command 139
.NET regular expressions
  URL 69
-wrap switch 281

A

AAD Sync
  about 53, 186
  deploying, in Lync resource forest 198-201
  rules 202-206
  synchronization services 202-206
  URL 198
Active Directory (AD) 57, 299
Active Directory Federation Services (AD FS) 34, 219
AD FS Relying Party Trust, working
  Claim 39
  Claims provider 40
  Relying party 40
  Relying party trust 40
  Trusted source 39
administrative rights
  controlling, with custom cmdlets 8, 9
  controlling, with RBAC 8, 9
Application Request Routing (ARR)
  about 26
  used, for configuring reverse proxy 26-32
app password
  adding, for mobile clients 45, 46
appUri parameter 22
A-records 299
assessment
  preparing 159

Audio/Video Multi-Point Conference Unit (AVMCU) 252
authentication protocols 34
auto attendants
  URL 105
Azure Active Directory Synchronization Services. *See* **AAD Sync**
Azure Portal
  URL 55

B

backend databases 290-292
Best Practices Analyzer (BPA)
  about 349
  testing 349-351
  URL 349
BYOD (Bring Your Own Device) 151

C

Call Admission Control (CAC)
  about 19, 160, 304
  enabling 324-327
call detail recording (CDR)
  URL 243
Call Diagnostic Reports 244-248
Call Flow
  investigating, with Snooper Flow Chart 340, 341
call forwarding
  controlling 87
Call Leg Media Quality Report 252, 253
Call Pickup Groups
  managing, Lync2013CallPickupManager 1.01
    used 143-145

catalog, of Lync devices
  URL 151
Centralized Logging Tool
  about 128
  URL 129
Central Management Store (CMS) 14, 61
certificates
  about 284-290
  managing, for desk-phones
      authentication 19-21
Certification Authority (CA) 19
challenges, users 149
Channel 9
  URL 61
Cisco Unified Communications
      Manager (CUCM) 252
client authentication logging
  troubleshooting with 62
clients
  Edge Server internal interface 327-329
  troubleshooting, with MLCA 359-361
cloud
  users, moving to 222-225
CLS
  about 128, 336, 342
  customizing, CLSController used 347, 348
CLS and configuration scenarios
  URL 132
CLSController
  used, for customizing CLS 347
cmdlets, Remote PowerShell for Lync Online
  URL 218
CNAME-records 299
codecs
  URL 166
command line
  tracing from, with OCSTracer 346, 347
commands
  used, for running script 121
commands, Response Group 139
conferencing security
  enhancing 18, 19
configuration
  backing up 270-273
Country Code (CC) 64

CsHealthMonitoringConfiguration
  URL 281
custom cmdlets
  administrative rights, controlling with 8, 9

D

data, network
  gathering 307-310
Demilitarized Zone (DMZ) 27
deployment
  verifying, with RCA 362-364
desk-phones authentication
  certificates, managing for 19-21
DHCP 21, 300
dial plans
  about 64-69
  defining 69-72
dial plans and voice routing
  URL 69
Differentiated Services
      Code Point (DSCP) 310
DigiCert Certificate Utility, for Windows
  URL 27
Directory Sync (DirSync)
  about 53
  prerequisites 48
  used, for authentication with online
      services 46-53
Distributed Denial-of-Service (DDoS) 21
Domain Name Server (DNS) 227, 299
DSCP tagging
  clients, preparing for 327-329
  servers, preparing for 327-329

E

Edge Server internal interface 330
Enable-CsComputer cmdlet 13
Enable-cstopology
  URL 270
end-user representatives
  requirements 149
Enterprise Voice
  about 63
  configuring 65

ethical walls
  applying, for federation security 23-25
Exchange
  and Lync 2013, integration level
    between 90, 91
  Outlook Web App mailbox policy,
    configuring 117
  OWA virtual directories, configuring for Instant
    Messaging 116
  web.config file, editing on client access
    servers 116, 117
Exchange 2013
  configuring, as partner applications 107
  configuring, as partner applications
    on Lync 2013 108
Exchange 2013 Outlook Web App
  Lync 2013, integrating with 114, 115
Exchange 2013 UCS
  using, via Lync 2013 configuration 113
Exchange backup 300
Exchange certificates
  enabling 96, 97
  Exchange UM dial plan, configuring 98-100
  Exchange UM dial plan, creating 98-100
  Exchange Util configuration script,
    running 100, 101
  Lync dial plan, creating 100
  Lync Unified Messaging configuration tool,
    running 101, 102
  requesting 94, 95
Exchange eDiscovery search
  URL 109
Exchange In-Place Hold
  URL 211
Exchange Management Shell
  UM-enabled users 104
  UM-enabled users, per mailbox policy 105
  Unified Messaging connectivity, testing 105
  users, disabled 104
  users, enabled 104
  using 95, 104
Exchange Online
  used, for Lync resource forest 184-186
Exchange UM dial plan
  configuring 98-100
  creating 98-100

Exchange Unified Messaging integration
  and Lync 2013 91
Exchange Util configuration script
  running 100, 101
Exchange version
  versus Lync version 90
Exchange Web Services (EWS)
  about 38
  features 91
Expresso 3.0
  URL 72
extended page tables (EPT) 157

F

failover
  configuring 83-85
federation security
  ethical walls, applying for 23-25
file servers 300
file services 296-298
filters
  URL 23
FIM
  about 53, 188
  configuring, in Lync resource forest 187-193
  forests, synchronizing with 194-197
  URL 193
firewall configurations 301
Flow Chart (Call Flow) 340
foreach command 290
forests
  synchronizing, with FIM 194-197
Forward Error Correction (FEC) 305
fully-functional voice configuration
  creating, Lync Dialing Rule Optimizer 123-127
Fully Qualified Domain Name (FQDN) 16

G

Get-CsOnlineUser cmdlet 218
Get-CsTenant cmdlet 216
Get-CsTenantHybridConfiguration cmdlet 216
Get-cstopology
  URL 270

Get-UnusedNumbers script
  about 138
  URL 138
  used, for managing phone numbers 138, 139
  using 139-143
group call pickup 143

H

heavy conference user 164
human factor 148
hybrid deployment
  configuring 220-222
  planning 220-222

I

Identity parameter 22
Import-PSSession cmdlet 215
Instant Messaging (IM) 23, 207, 304
Internal DNS 185
Internet Information Services (IIS) 27
Internet Telephony Service Provider (ITSP) 72
IP-public branch exchange (PBX) 82

L

Least Cost Routing (LCR) 76
load balancing
  configuring 83-86
Location-Based Routing (LBR) 76
Location Information (LIS) database
  about 279-281
  backing up 280
  restoring 280
log files
  examining, Snooper used 336-338
Lync
  about 148
  passive authentication, configuring for 35-41
Lync 2010 and 2013 Bandwidth Calculator
  URL 159
Lync 2010 Edge Servers
  URL 22
Lync 2013
  and Exchange, integration level
    between 90, 91
  and Exchange Unified Messaging
    integration 91
  configuring, as partner applications 107
  configuring, as partner applications on
    Exchange 2013 108
  configuring, to use Exchange 2013 for
    archiving 109
  configuring, to use Exchange 2013 UCS 113
  integrating, with Exchange 2013 Outlook
    Web App 114
  with SCOM 254-259
**Lync 2013 and 2013 bandwidth
  calculator tool**
  URL 171
Lync 2013 and Exchange 2013
  OAuth, configuring between 105-107
Lync2013CallPickupManager
  about 143
  URL 143
Lync2013CallPickupManager 1.01
  used, for managing Call Pickup
    Groups 143-145
Lync 2013 Centralized Logging Tool
  used, for tracing 128-131
**Lync 2013 configuration,
    using Exchange 2013**
  Exchange archiving, enabling 109
  ExchangeArchivingPolicy property per user,
    configuring 110
  Exchange archiving, to external
    communication 110
  Exchange archiving, to internal
    communication 110
  Lync, archiving on user Exchange
    mailbox 110
**Lync 2013 configuration, using Exchange
    2013 UCS**
  UCS settings, listing 113
  UCS settings, managing 113
Lync 2013 Ignite
  URL 78
Lync 2013 monitoring reports
  installing 234-239
Lync 2013 Resource Kit
  URL 176
Lync 2013 Server installation
  URL 123

Lync 2013 update page
  URL 298
Lync 2013, with Exchange 2013 Outlook Web App
  Exchange, configuring for IM integration with OWA 115
  trusted application pool, creating on Lync for OWA 115
Lync Admin Center (LAC)
  about 208
  administering with 208-211
Lync archiving policy
  changing 112
  listing 112
Lync Bandwidth Calculator post
  URL 307
Lync certificates
  resources 32
Lync Conference 2014
  URL 78
Lync databases
  hardening 14-17
Lync Dialing Rule Optimizer
  about 66, 124
  URL 124
  used, for creating fully-functional voice configuration 123-127
Lync Dialing Rule Optimizer, parameters
  3-Digit Area Code 125
  3-Digit Local Exchange 125
  Change rulename base 125
  Country 125
  External Access # 125
  Gateway type 125
  Simple ruleset 125
  Sip Trunk Connection 126
  Use Extensions 126
Lync dial plan
  creating 100
  URL 100
Lync hybrid deployment 218, 219
Lync information
  reviewing, with OCSLogger 342-346
Lync instructor-led training
  URL 153

Lync Management Shell
  used, for obtaining reports 112
Lync Online
  about 207, 208
  Windows Azure Directory, managing for 54-56
Lync Online cmdlets
  using 215-217
Lync Online issues
  debugging 226-231
Lync Online Remote PowerShell
  using 211-215
Lync Pilot Deployment Health Analysis Tool
  prerequisites 133
  used, for identifying recurrent issues 132-137
Lync PowerShell cmdlets
  URL 120
Lync prerequisites
  installing 120-122
Lync resource forest
  AAD Sync, deploying in 198-202
  Exchange Online, used for 184-187
  FIM, configuring in 187-193
Lync Server
  about 33, 329
  features 7
  hardening 10-13
  requirements 156
Lync Server 2010
  URL 20
Lync Server 2013
  about 263
  troubleshooting 335
  URL 8
Lync Server 2013 Debugging Tools
  URL 336
Lync Server backup 300
Lync Unified Messaging configuration tool
  running 101, 102
Lync users list, with Exchange archiving policy
  obtaining 112
Lync version
  versus Exchange version 90
Lync virtualization
  applying 155-159

M

many-to-many (M:N) trunk routing 78
media bypass 333
Media Quality Diagnostic Reports 248-251
MFCMAPI tool
  URL 110
Miami 170
Microsoft Forefront Identity Manager (FIM) 2010 R2 187
Microsoft Lync 2013 rollout and adoption success kit
  URL 153
Microsoft Lync 2013 Technet article
  URL 90
Microsoft Lync Connectivity Analyzer. *See* MLCA
Microsoft Lync Deployment Planning Services
  URL 154
Microsoft Lync Server 2013 147
Microsoft Office pages
  URL 24
Microsoft RASK
  URL 132
Microsoft SIP Processing Language (MSPL) 21
Microsoft SVVP supportability pages
  URL 156
Microsoft Web Platform Installer (MWPI)
  URL 26
MLCA
  clients, troubleshooting with 359-361
  URL 359
mobile clients
  app password, adding for 45, 46
mobile user 164
MOS
  URL 251
msExchMasterAccountSid attribute 175
Multi-Factor Authentication (MFA) 41, 43
multiple Lync identities
  switching between, Profiles for Lync (P4L) used 127

N

National Destination Code (NDC) 64
nested page tables (NPT) 157
network
  adding, to topology 315-320
  personas, defining for 162-164
  sites, defining for 164-166
Network Address Translation (NAT) 27
network bandwidth policies
  creating 312-314
network interfaces (NICs) 27
network readiness assessment 159-162
network traffic
  capturing, with Wireshark 351-359
non-uniform memory access (NUMA) 157
NT LAN Manager (NTLM) 34

O

OAuth
  configuring, between Lync 2013 and Exchange 2013 105, 106
  testing 108
OCSLogger
  about 336
  Lync information, reviewing with 342-346
OCSTracer
  command line, tracing with 346, 347
Office Web Apps 122
on-premises
  users, moving to 225
Open Authorization 2.0 (OAuth 2.0) 56
Organizational Unit (OU) 224
Outlook Voice Access (OVA)
  URL 92
Outlook Web App mailbox policy
  configuring 117
OWA virtual directories
  configuring, for Instant Messaging 116

P

parameters, Lync Pilot Deployment Health Analysis Tool
  Customer Name 134
  Monitoring Server URL 134
  SQL Server 133
  User ID 133
  Use Windows Security 133
partner applications
  testing 108

passive authentication
  configuring, for Lync 35-41
Peer to Peer (P2P) 304
Persistent Chat database 277, 278
personas
  defining, for network 162-164
port ranges
  controlling, for traffic 330, 331
PowerShell engine 119
Privacy Relationships 24
Private Branch Exchange (PBX) 66
Profiles for Lync (P4L)
  URL 127
  used, for switching between multiple Lync identities 127
PSTN usage, Location-Based Routing (LBR)
  configuring 76, 77
PSTN usage, voice policy
  configuring 72-76
Public DNS 185
public key infrastructure (PKI) 300
Public Switched Telephone Network (PSTN) 64
Publish-cstopology
  URL 270

Q

QoEMetrics 238
Quality of Experience (QoE) 238, 243, 305
Quality of Service (QoS) 19, 150, 160, 304

R

rationale
  heavy conference user 164
  mobile user 164
  reviewing 164
  standard user 164
recovery plan
  about 299
  Active Directory 299
  DHCP 300
  DNS 299
  Exchange backup 300
  file servers 300
  firewall configurations 301
  Lync Server backup 300
  miscellaneous 301
  public key infrastructure (PKI) 300
  reverse proxy 301
  router configurations 301
  SQL Server 300
  switch configuration 301
recurrent issues
  identifying, Lync Pilot Deployment Health Analysis Tool used 132-137
regions links
  creating 320-324
Regular Expressions (RegEx) 123, 127
Reliability Report tab 135
Remote Connectivity Analyzer (RCA)
  deployment, verifying with 362-364
  URL 227
report
  selecting 239-243
requirements, users
  gathering 153, 154
resource forest
  planning 174-183
resource forest topology 174
resource monitoring 158
Response Group Services
  configuring 282, 283
results
  analyzing 167-171
  reviewing 167-171
reverse proxy
  about 301
  configuring, Application Request Routing used 26-32
RFC 3966
  URL 127
Role-Based Access Control (RBAC)
  about 82, 182
  administrative rights, controlling with 8, 9
router configurations 301
routes
  creating 320-324
  enabling 78-81
RTCUniversalServerAdmins 279, 312, 315
Run All Queries feature 134

S

SCOM
   Lync 2013, used with 254-259
   URL 254
Search-LineURI
   about 138
   URL 138
   used, for managing phone numbers 138, 139
Secondary Extension Feature Activation Utility (SEFAUtil) 87
secure Lync Edge
   deploying 21, 22
Security Configuration Wizard (SCW) 10
SEFAutil tool 143
servers
   Edge Server internal interface 330
   Lync Servers 329
   preparing, for DSCP tagging 327-329
server-to-server authentication
   configuring 56-61
   scenarios 57
server-to-server authentication certificate, on Lync 2013
   URL 106
service connection point (SCP) 106
Session Border Controller (SBC) 82
Set-Cs2013Features
   used, for installing Lync prerequisites 120-122
settings, Unified Messaging
   changing, on user 103
settings, voice 292-296
single pool deployment
   URL 83
Single-root I/O virtualization (SR-IOV) 157
sites
   defining, for network 164-166
Snooper
   about 336
   URL 132
   used, for examining log files 336-338
Snooper Flow Chart
   Call Flow, investigating with 340, 341
SQL Server 300

SQL Server 2008 R2 Security Best Practices
   URL 18
SQL Server Express installation 122
SQL Server Reporting Services (SSRS) 239
SRV-records 299
standard user 164
Subscriber Number (SN) 64
switch configuration 301
switches
   - Level 278
   - Scope 278
   - Startdate 279
synthetic transactions
   configuring 260-262
System Center 2012 R2 Operations Manager. *See* **SCOM**

T

tabs, Network Configuration
   about 306
   Bandwidth Policy 306
   Global 306
   Region 306
   Region Link 306
   Region Route 306
   Site 306
   Subnet 306
TechNet post Manage Trusted Root Certificates
   URL 185
TechNet wiki
   URL 17
tools
   features 120
topology
   backing up 264-269
   networks, adding to 315-320
Topology Builder 264
tracing
   Lync 2013 Centralized Logging Tool, used for 128-131
traffic
   port ranges, controlling for 330, 331
Transformations option 205

Transparent Data Encryption (TDE) 17
trunks
  validating 82
two-factor authentication
  enabling 41-44

U

UCMA runtime 4.0
  URL 228
UCS feature
  testing 114
UCS settings
  user services policy, changing 113
  users, with UCS enabled 113
UDC agent 238
UK-International rule 71
UK-Premium rule 71
UK-TollFree rule 71
Unified Communications
    Managed API (UCMA)
  URL 114
Unified Contact Store (UCS) 91, 112
Unified Messaging
  about 91
  features 91, 92
  settings, changing on user 103
  users, enabling for 102, 103
Unified Messaging, features
  auto attendant 92
  Call Answering 92
  Outlook Voice Access (OVA) 92
Unified Messaging integration
  configuring 93-105
  Exchange certificates, configuring for 93
  prerequisites 93
  users, managing 102
user database 274-276
User Principal Name (UPN) 49
users
  moving, to cloud 222-225
  moving, to on-premises 225, 226

users' expectations
  meeting 148-151
users, Unified Messaging
  enabling 102, 103
  managing 102
user training
  about 151-153
  benefits 152

V

Virtual Machine Queue (VMQ) 157
voice dial plans 292-296
Voice over IP legislation
  URL 151
Voice over IP (VoIP) 150
voice policies 292-296
voice routing 64-69

W

watcher node
  configuring 260, 261
Web Application Proxy (WAP) 26
web.config file
  editing, on client access servers 116, 117
Windows Azure Active Directory (WAAD) 53
Windows Azure Directory
  managing, for Lync Online 54-56
Windows Identity Foundation (WIF) 260
Windows PowerShell Module for Lync Online
  URL 212
Wireshark
  about 351
  network traffic, capturing with 351-359
  URL 351

Thank you for buying
Lync Server Cookbook

About Packt Publishing

Packt, pronounced 'packed', published its first book, *Mastering phpMyAdmin for Effective MySQL Management*, in April 2004, and subsequently continued to specialize in publishing highly focused books on specific technologies and solutions.

Our books and publications share the experiences of your fellow IT professionals in adapting and customizing today's systems, applications, and frameworks. Our solution-based books give you the knowledge and power to customize the software and technologies you're using to get the job done. Packt books are more specific and less general than the IT books you have seen in the past. Our unique business model allows us to bring you more focused information, giving you more of what you need to know, and less of what you don't.

Packt is a modern yet unique publishing company that focuses on producing quality, cutting-edge books for communities of developers, administrators, and newbies alike. For more information, please visit our website at www.PacktPub.com.

About Packt Enterprise

In 2010, Packt launched two new brands, Packt Enterprise and Packt Open Source, in order to continue its focus on specialization. This book is part of the Packt Enterprise brand, home to books published on enterprise software – software created by major vendors, including (but not limited to) IBM, Microsoft, and Oracle, often for use in other corporations. Its titles will offer information relevant to a range of users of this software, including administrators, developers, architects, and end users.

Writing for Packt

We welcome all inquiries from people who are interested in authoring. Book proposals should be sent to author@packtpub.com. If your book idea is still at an early stage and you would like to discuss it first before writing a formal book proposal, then please contact us; one of our commissioning editors will get in touch with you.

We're not just looking for published authors; if you have strong technical skills but no writing experience, our experienced editors can help you develop a writing career, or simply get some additional reward for your expertise.

Getting Started with Microsoft Lync Server 2013

ISBN: 978-1-78217-993-1   Paperback: 122 pages

Everything you need for understanding and working with Lync 2013 in a fast-paced manner

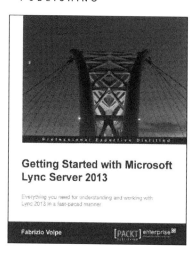

1. Understand and deliver the top required features such as Enterprise Voice, Persistent Chat, and mobility with step-by-step guides.
2. Deploy a working environment applying load balancing and fault tolerant solutions.
3. Create a collaborative space around the user's needs, containing all the information and document history using the Persistent Chat Server.

Microsoft Lync 2013 Unified Communications: From Telephony to Real Time Communication in the Digital Age

ISBN: 978-1-84968-506-1   Paperback: 224 pages

Complete coverage of all topics for a unified communications strategy

1. A real business case and example project showing you how you can optimize costs and improve your competitive advantage with a Unified Communications project.
2. The book combines both business and the latest relevant technical information so it is a great reference for business stakeholders, IT decision makers, and UC technical experts.

Please check **www.PacktPub.com** for information on our titles

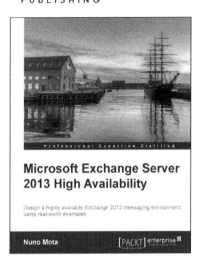

Microsoft Exchange Server 2013 High Availability

ISBN: 978-1-78217-150-8　　　Paperback: 266 pages

Design a highly available Exchange 2013 messaging environment using real-world examples

1. Use the easy-to-follow guidelines and tips to achieve the highest availability.
2. Covers all the aspects that need to be considered before, during and after implementation of high availability.
3. Packed with clear diagrams and scenarios that simplify the application of high availability concepts such as site resilience.

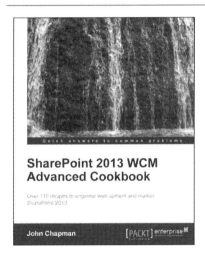

SharePoint 2013 WCM Advanced Cookbook

ISBN: 978-1-84968-658-7　　　Paperback: 436 pages

Over 110 recipes to engineer web content and master SharePoint 2013

1. Brand SharePoint with master pages and page layouts.
2. Catalog content with cross site publishing.
3. Enhance the user experience with custom controls.
4. Create multilingual sites.

Please check www.PacktPub.com for information on our titles